Learning and Analytics in Intelligent Systems

Volume 27

The main aim of the series is to make available a publication of books in hard copy form and soft copy form on all aspects of learning, analytics and advanced intelligent systems and related technologies. The mentioned disciplines are strongly related and complement one another significantly. Thus, the series encourages cross-fertilization highlighting research and knowledge of common interest. The series allows a unified/integrated approach to themes and topics in these scientific disciplines which will result in significant cross-fertilization and research dissemination. To maximize dissemination of research results and knowledge in these disciplines, the series publishes edited books, monographs, handbooks, textbooks and conference proceedings.

More information about this series at https://link.springer.com/bookseries/16172

Panagiotis T. Artikis · Constantinos T. Artikis

Random Contractions in Global Risk Governance

Springer

Panagiotis T. Artikis
Department of Accounting and Finance
University of West Attica
Egaleo, Greece

Constantinos T. Artikis
Department of Tourism
Ionian University
Corfu, Greece

ISSN 2662-3447 ISSN 2662-3455 (electronic)
Learning and Analytics in Intelligent Systems
ISBN 978-3-030-95690-5 ISBN 978-3-030-95691-2 (eBook)
https://doi.org/10.1007/978-3-030-95691-2

This Springer imprint is published by the registered company Springer Nature Switzerland AG
The registered company address is: Gewerbestrasse 11, 6330 Cham, Switzerland

Series Editor's Foreword

Undoubtedly, the advances in the information, communication, transport, and energy technologies have led the World into strong interconnection and interdependence at the economic, social, and geopolitical level. At the same time, humanity is now facing challenges and threats (e.g., security, military, health, or environmental threats, but also threats from advances in technologies such as Artificial Intelligence and Data Science) on a global/worldwide scale, which can be addressed only via collaboration among countries and other stakeholders on a matching global/worldwide scale.

Consequently, suitable tools for decision-making need to be made available, which can predict the evolution of complex systems reliably and in sufficient depth. Stochastic modeling research has been devising methodologies that show promise to be particularly useful in the hands of decision-makers, as they have proven to be efficient in the modeling and prediction of complex systems.

In this monograph, Profs. Panagiotis T. Artikis and Constantinos T. Artikis focus on a specific class of stochastic models of complex systems, known as *Random Contractions*. Random Contractions refer to stochastic multiplicative models which are useful for describing, analyzing, and solving significant problems across many scientific disciplines. In more detail, the authors start with a discussion of global risks and their corresponding stochastic modeling and proceed with rigorous formulation of stochastic models and their random contraction representation. At this point, Profs. Artikis and Artikis also introduce eight new random contractions. At the final part of the book, they apply and interpret the random contraction model to global risk governance.

Overall, the book is very well written, starting from introductory concepts and leading the reader to the most advanced and most recent research results in this area. As it is, the book is addressed not only to the expert in the fields of stochastic modeling and global risk governance, but it also appeals to the newcomer who wishes to learn about these fields. Moreover, the book will attract the interest of readers from other areas as well who wish to get versed in these significant scientific disciplines.

As Editor of the *Learning and Analytics in Intelligent Systems* series of Springer, I am happily presenting this book to the research communities worldwide. I congratulate the authors for their great work, in confidence that their book will help its readers understand the significant research areas of stochastic modeling and random contractions. I am also confident that the approaches and methodologies included in this book will find application in many scientific areas and provide solutions to diverse problems. Finally, I encourage the authors to continue their research work in this area and, perhaps, update the scientific communities with future books of theirs.

December 2021

George A. Tsihrintzis
Department of Informatics
University of Piraeus
Piraeus, Greece

Preface

Global governance gathers a large number of different performing actors at a global level. The main objective of global governance is to provide frameworks, measures, and guidance in global economic markets, social structures, security, and peace. The expansion of organized global markets, development of global communication networks, and the vast number of global organizations has expanded the interconnection among countries by bringing people, groups, and societies closer in a time period with issues and threats call for worldwide collaboration.

A large number of actors characterize and shape the structure of global governance. Current forms of global governance must support adaptability over inflexibility and allow rise of unused activities and concepts.

The formulation, investigation, and application of stochastic models constitute research activities of exceptional theoretical and practical importance, concerning the future behavior and evolution of real-world systems. Stochastic models are considered as suitable tools for the development and the successful application of diverse optimal control plans and the proper instruction for decision-making process.

It is acceptable that stochastic models constitute an extremely important category of models having an exceptional contribution to the precise description, the thorough investigation, and the good performance of many complex real-world systems arising in a wide variety of fast-growing disciplines with particular practical importance.

In particular, random contractions are generally recognized as strong analytical tools of probability theory with significant practical applications in a wide variety of disciplines.

The powerful probabilistic role and the mathematical form of random contractions constitute the main advantages for considering a random contraction as a stochastic model of particular applicability in various important theoretical and practical disciplines.

Random contractions constitute a class of stochastic multiplicative models which comprise a strong analytical tool providing analysts, modelers, decision-makers, and other experts with exceptionally valuable probabilistic information for effectively treating many worldwide complex systems.

The authors would like to thank Profs.-Drs. Maria Virvou and George A. Tsihrintzis of the University of Piraeus, Piraeus, Greece, for their encouragement, fruitful discussions, and comments over the last few years, which have helped to shape this monograph.

Egaleo, Greece Panagiotis T. Artikis
Corfu, Greece Constantinos T. Artikis

Contents

1 Stochastic Concepts in Global Risk Governance Operations 1
1.1 Introduction 1
1.2 Structural Aspects of Global Risks 3
1.3 Concepts of Global Risk Governance and Probability Theory 7
1.4 Information Dimensions of Global Risk Governance 12
1.5 Conclusions 24
References 24

2 Consideration in Modeling of Random Contractions 27
2.1 Introduction 27
2.2 Defining Random Contractions 33
2.3 Distribution Function of a Random Contraction 35
2.4 Probability Density Function of a Random Contraction 36
2.5 Characteristic Function of a Random Contraction 37
2.6 Several Fundamental Classes of Random Contractions 41
2.7 Class of a-Unimodal Random Contractions 42
2.8 Class of υ-Unimodal Random Contractions 46
2.9 Class of Random Contractions of Maxima of Random Number of Random Variables 50
2.10 Class of Random Contractions of Minima of Random Number of Random Variables 58
2.11 Class of Random Contractions of Present Value Stochastic Integrals 65
2.12 Class of Random Contractions of Random Sums 76
2.13 Class of Random Contractions of Present Values 90
2.14 Class of Random Contractions of Bernoulli Selections 94
2.15 Conclusions 102
References 102

3 Formulations and Investigations of Random Contractions 105
3.1 Introduction 105
3.2 Formulation of a Stochastic Multiplicative Model 108
3.3 Representation as a Random Contraction of a Stochastic Multiplicative Model 109
3.4 Extending the Contraction of a Random Sum 122
3.5 Interpretation of the Formulated Random Contraction in Stochastic Discounting 130
3.6 Characteristic Function of the Formulated Stochastic Discounting Model 136
3.7 Extension of a Stochastic Multiplicative Model 140
3.8 Representation as a Random Contraction of an Extension of a Stochastic Multiplicative Model 142
3.9 Characteristic Function of an Extension of a Random Contraction 153
3.10 Characterizing an Extension of a Random Contraction 162
3.11 Representation of an Extension of a Random Contraction as a Stochastic Discounting Model 168
3.12 Characteristic Function of the Representation as a Stochastic Discounting Model of an Extension of a Random Contraction 175
3.13 Characterizing the Representation as a Stochastic Discounting Model of an Extension of a Random Contraction 179
3.14 Conclusions 186
References 187

4 Interpretations of Random Contractions in Global Risk Governance 189
4.1 Introduction 189
4.2 Number of Impaired Systems 194
4.3 Global Risk Severity 195
4.4 Discounted Global Risk Severity 197
4.5 Global Risk Severity Reduction 198
4.6 Global or Regional Risk Severity 199
4.7 Discounted Global or Regional Risk Severity 201
4.8 Global or Regional Risk Severity Reduction 202
4.9 Practical Research Purposes 203
4.10 Model of Impaired Systems Number 211
4.11 Model of Global Risk Severity 217
4.12 Model of Discounted Global Risk Severity 227
4.13 Model of Global Risk Severity Reduction 237
4.14 Model of Global or Regional Risk 247
4.15 Model of Discounted Global or Regional Risk Severity 256
4.16 Model of Global or Regional Risk Severity Reduction 269
4.17 Conclusions 283
References 284

Chapter 1
Stochastic Concepts in Global Risk Governance Operations

Abstract The formulation of stochastic models suitable for proactive treatment of global information risks incorporates structural concepts of global information risk governance and fundamental results of probability theory. The theoretical and practical contribution of the first chapter of this work is comprised of the above mentioned structural concepts and fundamental results. As a result, the contribution of formulating stochastic models aimed at improving the discipline of global risk governance is easily accepted as particularly valuable in theory and practice.

Keywords Global risk · Global risk governance · Information science · Probability theory · Stochastic model

1.1 Introduction

In this work global risks are recognized as incorporating very significant provision of economic and social support with a multistakeholder support, global geographical dimension, and inter-industrial interest. In addition, global risks demand an extremely extensive priority in decision making allowance. It constitutes a general acceptance that the concept of global risk has not been defined quite clearly. More precisely, the concept of global risk is incorporated for the vague description of a risk whose the occurrence might cause damages on a global level. From the fact that the ongoing risks and forthcoming risks constitute two different types of risks but the adoption of global risks as a new scientific discipline is quite readily recognized as a particularly important theoretical and practical activity. It is of exceptional importance to mention that a wide variety of methodological, conceptual, and cultural issues appear in the area of global risks. It is easily adopted that the consideration of global risks as a completely separate scientific discipline is based on several very significant practical reasons. In addition, information is not usually available and implementations of various global risk severity reduction operations are frequently very expensive. Moreover, decision making on designating activities and resources are completely necessary for obtaining extremely useful deductions. This work can be seen as a contribution to

P. T. Artikis and C. T. Artikis, *Random Contractions in Global Risk Governance*, Learning and Analytics in Intelligent Systems 27,
https://doi.org/10.1007/978-3-030-95691-2_1

the area of ongoing global risks and the area of forthcoming global risks, particularly necessary for the implementation of very important interdisciplinary research activities. The theoretical and practical results of this work are strongly connected to several quite useful factors. More precisely, researchers working on one of these types of global risks will have the opportunity to obtain experience about the other types of global risks [47]. Moreover, practitioners from different disciplines will have the opportunity to consider and treat various global risks for a broader aspect. Analysts and practitioners incorporated in strongly supporting effective policies can greatly increase their practical and theoretical abilities by increasing their connections with the state of the art in undertaking new and extensive research activities on global risks. The present work mainly concentrates on undertaking further research activities and active public incorporation on global risks for the development and the implementation of strategies for reducing the severity of forthcoming such risks [14, 24]. The presence of interdisciplinary groups of researchers and the presence of groups of practitioners about global risks will substantially contribute to find satisfactory solutions for the various problems of the present century. In other words the introduction of concepts and methods from other disciplines to the discipline of global risks is easily seen that such an introduction can be particularly useful for the discipline of global risk governance which is readily adopted as a very significant theoretical and practical extension of the discipline of global risks. More precisely, the present work can be considered as an attempt of introducing several fundamental concepts of probability theory in some structural operations of global risk governance [14, 25].

It is completely obvious that additional research activities on fundamental operations of global risk governance are considered as exceptionally valuable if they provide concentration on the formulation and implementation of stochastic models for measurement and treatment of various global risks, creation and use of methodologies for obtaining information needed for making forecasts on the occurrences of global risks, development of a particularly efficient interdisciplinary and international group of global risk experts with theoretical and practical backgrounds being able for constructing and implementing operations and strategies for reducing the severity of various global risks within the extensively wide frame of challenges and opportunities for developing a very substantial protection of humanity [20, 40, 42].

Since global risks are not incorporated to national borders it is easily adopted that such risks cannot be managed through the activities of a single sector. Moreover, global risk governance requires an integrative consideration involving governments, intergovernmental organizations, industry and civil society [20]. It is readily accepted the particularly important contribution of the extensive knowledge and proactive management of forthcoming global risks that strongly affect the human health and safety, the government, the economy and the society as a complex system. Moreover, it is easily seen that the above mentioned contribution involves consideration of structural concepts of global risk governance, confrontation of strong problems of global risks, and development of global risk governance policy for leading decision makers. It is generally recognized that very important new contributions to global risk governance are very valuable if optimal risk decisions are to be taken and to support confidence to activities and structures incorporated in making of such decisions. The structure of the international risk governance council provides a very strong tool for the investigation, communication and management of global risks by including useful scientific results, global risk perception, social issues and social principles. This kind of structure strongly supports to develop a very good and combined consideration of global risk governance and consequently facilitates decision makers with developing the proper questions and creating valuable global risk governance strategies for the increasingly complex opportunities and global risks arising in the present century [47].

It is very well known that the scientific disciplines of strategic thinking, global risk governance, and stochastic modeling are generally recognized as very important from a theoretical and a practical point of view [10]. In addition, the scientific discipline of stochastic modeling constitutes a valuable analytical tool for the discipline of strategic thinking. Clearly the formulation, investigation, and implementation of various stochastic models substantially facilitate fundamental activities in the discipline of strategic thinking [11, 56]. More clearly, the discipline of stochastic modeling provides leading analysts in strategic thinking with the ability to describe and manipulate random factors in a systemic way. Consequently, it is easily seen that the theoretical and practical interconnections between the discipline of stochastic modeling and the discipline of global risk governance constitute a very interesting scientific area for undertaking extensive and thorough research. The following sections of this chapter concentrate on structural concepts of global risks and concepts of global risk governance [15].

1.2 Structural Aspects of Global Risks

It is generally recognized that interdependency, heurist bias, and policy mistake are incorporated in the class of fundamental structural aspects of the concept of global risk. The present section concentrates on providing a brief exploration of these structural aspects. More precisely, this section concentrates in a thorough understanding of the nature of interdependency and offer an interpretation of how interdependency influences the activity in which global risks are managed and reduced. Moreover, this section reflects a human approach to global risk in general. Frequently, humans act with incomplete information by making extensive use of heuristics. In addition, policy mistakes are frequently connected to heurist biases. Consequently, such connections make quite clear how a policy, often intended to provide reduction to a global risk, can actually exacerbate that global risk [16].

It is easily understood that the support of firms for undertaking investments in global risk reduction operations, in a environment incorporating a particularly increasing number of interdependencies, constitute an extremely significant activity for policy makers. During the last two decades, researchers and practitioners are strongly concentrated on the thorough investigation problems arising in various areas of interdependent security. Adoption of proactive global risk strategies can reduce the damages of a country, firm or individual but the rejection of proactive global risk strategies from other countries, firms or individuals can result to more severe global damages [15, 65]. Consequently, the economic uncective of a decision maker to undertake investments for developing and implementing proactive global risk operations strongly depends on whether others are expected to undertake and implement such investment procedures. Clearly, the fact that a global risk frequently determined in part by the behavior of others gives a very complex structure to the incectives that countries, firms or individuals face to reduce or invest in global risk reduction operations [16]. In a very wide variety of security situations, the opinion of an actor that other actors will not undertake investment activities for supporting increase of security then it is obvious that the incective to do so will face a substantial reduction. In consequence, the final result may be all the actors avoid to undertake investment activities for improving protection. In such a case all the actors probably find themselves in a more difficult situation of security. On the contrary, if each actor accepts that all the other actors will also accept to undertake investment activities for supporting increase of security then the development and implementation of a strategy for proactive treatment of forthcoming global risks will be recognized as an optimal one. It is quite obvious that the development and implementation of proactive optimal strategies for treatment of global risks constitute the most constituent factor and valuable purpose of global risk governance. In addition, the advancement of the discipline of global risk governance is mainly based on the selection, investigation, and the implementation of operations for the successful treatment of forthcoming global risks. The present work mainly concentrates on the contribution of probability theory in the advancement of the practical applicability of global risk governance [16]. More precisely, the work introduces structural stochastic concepts and fundamental stochastic results for the formulation of stochastic models of various global risk governance operations particularly valuable in proactive treatment of several groups of global risks. Such stochastic models are mainly incorporated in facilitating the development of strategies for global risks. It is easily seen that strategies for forthcoming global risks are of exceptional practical importance but the development of such strategies is frequently extremely difficult. In this case, it is advisable the development of strategies for ongoing global risks. In general, the formulation of stochastic discounting models is suitable for the development of strategies for forthcoming global risks and the formulation of stochastic compounding models is suitable for the development of strategies for ongoing risks. The main advantage of stochastic discounting models in the development of strategies for forthcoming global risks consists of providing the ability to analysts and leading decision makers to think and act in a proactive way. If policy consideration is realized under conditions of interdependency then it is also obvious that a significant part of cost is undertaken by competitors and other customers.

Heuristic is generally recognized as an increasingly complex and uncertain word. Humans are not able to make fully informed decisions by making use of imperfect information. Moreover, human decisions very frequently are based on approximations of the real world or equivalently on shortcuts that facilitates the making of quite fast decisions by resorting to learned behaviors. These approximations of the real world and the rules of thumbs implementations are quite well known as heuristics. The development and implantation of heuristics is very strongly connected with the fact that human brains have been specialized in very fast decision making at the expense of processing complexity. In the original condition of humanity, survival was based upon the very fast description, analysis and effective treatment of an ongoing or forthcoming global risk. Frequently heuristics is particularly valuable in faster and easier decision making. It is easily understood the heuristics can also lead to false judgments, particularly in global risk perception [62].

The advantage of researchers and practitioners to substantially implement effective assessment of various global risks can be reduced be several heuristic biases. The majority of such heuristic biases can exacerbate the effects of other biases. It is quite obvious that the incorporations of significantly diversified networks can strongly support the avoidance of many of the heuristic biases. Clearly, the incorporation of a group of events by making use of a similar one, concentration on situations which seem to support the original conditions of researchers and practitioners rather to support the assumptions of specialists who have doubts about the original conditions, overestimation of the human abilities to asses clinging mentally events or figures heard in a particular context [2, 3].

From a historical point of view, policy in financial markets has the ability to reduce and exacerbate various global risks. More precisely, the understanding of global risks is difficult and it is dependent on a wide variety of interrelated factors. In addition, global risks are strongly influenced by policy errors, whether on the part of governments, regulators or central banks. In several situations, the implementation of policy has dampened the influences of the market and thereby decreased volatility. Moreover, in other situations the implementation of policy has significantly exacerbated risk. During the last decade, mistakes of regulatory or monetary policy have either restricted the strength of the market to reduce risks or exacerbate risks within the market itself. Policy mistakes are adopted as quite recognizable in retrospect but very rarely recognizable at the time point of decision making by governments and regulators on judgment, experience, incomplete information, and the balancing of alternative ways of actions. It is easily understood that the availability of information about the presence of mistakes in policy can provide the most valuable reduction for the avoidance mistakes in the future. It is readily accepted that risk identification and risk assessment constitute the operations for the reduction of global risks. Moreover, it is quite easily understood that the assessment of individual global risk is extremely simpler than the reduction of global risks under conditions of uncertainty, interdependency and competing purposes. It is well known that global risk reduction incorporates two main completely different operations. More clearly, ongoing risk occurrences require the use of reactive global risk reduction operations and forthcoming global risk occurrences require the use of proactive global risk reduction operations. From a theoretical and practical point of view the proactive global risk reduction operations are easily recognized as more valuable for leading decision making than the reactive global risk reduction operations. It is particularly important to mention that the use of proactive global risk reduction operations are based on the availability of reliable information. The selection and the implantation of an operation for the reduction of an individual global risk will depend on prioritization, resources, and understanding. Some global risks can significantly be reduced with low costs by modifying ways of thinking and adopting new methods of action. Other global risks can significantly be reduced by requiring developments and implementations of many and very strong institutional activities. It is readily accepted that common elements of ongoing or forthcoming global risk reduction operations are alignment on priorities, common understanding, and identical activities to describe and face very difficult situations of quite collective cooperation. Reactive global risk reduction operations and proactive global operations are not mutually exclusive. Global risk governance strategies very frequently incorporate reactive and proactive global risk reduction operations. From the fact that it is not possible the prediction and prevention of all global risk occurrences then increase of resilience in a global level. It is readily understood that improving insight, enhancing information flow, refocusing incectives, improving investment and implementing through institutions are five extremely important structural elements for the construction and implementation of global risk severity reduction strategies [21]. More precisely, the definitions of the above five elements are quite sufficient for understanding the exceptional contributions of these elements to the development and implementation of an effective global risk governance strategy [49].

Prioritization among global risks constitutes a completely necessary activity for decision making about the best use of resources for global risk severity reduction operations. It is obvious that such resources are not unlimited and the prioritization of global risks requires the implementation of two fundamental activities. The first activity is the evaluation of the probability of occurrence of a global risk and the second is the evaluation of the severity of this global risk occurrence. More precisely, the first activity constitutes the evaluation of the probability function of the frequency of the global risk for a given time interval and the second activity constitutes the evaluation of the distribution function of the severity of the global risk for the same time interval. The above mentioned activities make quite clear that many structural concepts and methods of probability theory can substantially contribute to the development, investigation, and implementation of stochastic models strongly supporting the advancement of global risk governance operations of valuable theoretical and practical importance. The present work can be considered as an attempt to provide interpretations of very well known results of probability theory as important analytical tools facilitating the understanding and applicability of activities arising in the discipline of global risk governance. In consequence, this work concentrates on making quite clear that the discipline of global risk governance can be advanced by making use of stochastic models having extensive use in economics, risk management, operational research, cindynics, informatics and systemics [9, 12]. Concepts and methods of probability theory, particularly useful for formulation of stochastic models with valuable applications in developing and implementing global risk governance operations are incorporated in the following section. The main factors of global risk severity are not simple at the global level. First, the occurrences of combinations of global risks can influence all firms, even if global risks are frequently considered by individual firms and countries as exogenous. Second, global risks can create influences outside the main estimations of risk managers. Third, global risk occurrences can arise in extremely long time intervals. This fact makes necessary the realizations of comparisons between global risk severity operations over different generations in order to introduce objectivity and reliability in practice. Fourth, interconnections between global risk complicate prioritization and hence interconnections are easily adopted as particular structural elements in the global environment. Finally, global risk severity strongly demands the very strong cooperation of quite different groups. Consequently, the majority of global risks cannot be reduced by an individual firm.

1.3 Concepts of Global Risk Governance and Probability Theory

It is generally recognized that the scientific discipline of global risk governance mainly uses qualitative concepts. In other words, the establishment, advancement, and implementation of new results in global risk governance are very strongly connected with the concept of quality. However, the concept of quality is frequently combined with the concept of quantity in various situations arising in investigation and treatment of global risks. Moreover, it is of exceptional interest to mention that the presence of the concept of quantity in various areas of the global risk governance is usually connected with the concept of randomness. The direct connection of randomness with the theory of probability leads to conclusion that the discipline of global risk governance has a strong connection with the discipline of probability theory. From a theoretical point of view it follows that researchers in the discipline of probability theory can find new areas for undertaking probabilistic research activities concentrating in concepts and methods arising in the discipline of global risk governance. In addition, researchers in pure probability theory can establish new practical results by the interpretation of their theoretical results as tools for describing and investigating global risk governance operations. Furthermore, probabilists can contribute to the whole discipline of global risk governance by the formulation, investigation and implementation of stochastic models for developing of optimal global risk strategies for treating global risks. From the fact that proactivity is generally adopted as the most structural element of a proactive treatment of various global risks it follows that stochastic discounting models are quite suitable analytical tools for supporting leading decision makers in strongly contributing the choice of optimal global risk treatments. Since, in general the implementation of global risk treatment operation requires a very high financing then the implementation of such an operation is easily understood as an extremely important action. It is obvious that the presence of proactivity in such an operation then forthcoming global risks can be treated in a very effective and reliable way. In this case, the cooperation of experts in stochastic discounting modeling with experts in the implementation of proactive global risk treatment make quite clear the advantage of combing a qualitative scientific discipline with a quantitative scientific discipline [60].

Below, we consider practical situations which can be described and investigated by making use of methods which arise as combinations of global risk governance operations and results of probability theory. First, the severity of a global is readily recognized as a positive random value whose probabilistic behavior is described by the corresponding distribution function [50]. Similarly the severity of a random number of independent global risks is also readily recognized as the sum of a random number of independent and equally distributed positive random variables, or equivalently as a random sum of positive random variables, whose probabilistic behavior is described by the corresponding characteristic function [41, 54, 61]. The frequency of a global risk in a given time interval is obviously denoted by a discrete random variable whose probabilistic behavior is described by the corresponding probability generating function [32, 46]. In addition, the duration of a global risk is easily seen that it can be represented by a positive random variable whose probabilistic behavior can be described by the corresponding distribution function [4]. Similarly, the occurrence time of a global risk is represented by a positive random variable whose probabilistic behavior can be described by the corresponding distribution function [53]. The number of ongoing global risk occurrences at the present, called the time point zero, can be represented by a discrete random variable whose probabilistic behavior can be described by the corresponding probability generating function. Similarly, the number of forthcoming global risk occurrences at some given future time point can be represented by a discrete random variable whose probabilistic behavior can be described by the corresponding probability generating function. The time between two successive occurrences of a global risk can be represented by a positive random variable whose probabilistic behavior can be described by the distribution function corresponding to the first occurrence of the global risk. If a group of global risks are catastrophic then the research interests concentrates on the time of the first occurrence of a global risk. In this case we consider the global catastrophic risks from the point of view of competing global risks. In other words the interest on global catastrophic risks is replaced by the interest on competing global risks. Consequently, the research activities concentrates on the time of occurrence of the first competing global risk and the corresponding distribution function. In conclusion, the treatment of a group of global catastrophic risks can be transformed as the treatment of a group of competing global risks. Such a transformation is very important from a theoretical and practical point of view. More precisely, the theoretical importance of that transformation is based on the fact that the distribution function of the minimum of a random number of positive random variables is known [5, 6, 18, 26, 27]. This distribution function provides experts with extremely valuable information for leading decision making. It is quite obvious that the above mentioned positive random variables denote occurrence times of competing global catastrophic risks. In consequence, a probabilistic result significantly reduces a difficult theoretical problem to an easy one. The practical importance of the above transformation is based on considering a problem of global catastrophic risks as a problem of competing global risks. Furthermore, a result of pure probability theory combined with a concept of global risk governance provides practical experts with suitable and extremely valuable tools for providing relatively easy solutions to difficult problems arising in the treatment of global catastrophic risks.

In general, the minimum of a random number of positive independent random variables constitutes a result of pure probability theory with several interesting interpretations in global risk governance operations. We suppose that the positive random variables denote the severities of a random number of competing global risks. In this case, the minimum of a random number of independent positive random variables represents the best scenario for the evolution of the corresponding competing global risks. In consequence, the combination of a result of pure probability theory with a concept of global risk governance substantially contributes to the description and solution of a problem arising in the area of competing global risks. In conclusion, the presence of the minimum of a random number of positive independent random variables in the discipline of global risk governance makes quite clear that such a minimum can be introduced as a principal component in various stochastic models with valuable contribution in the description, investigation, and implementation of very important global risk operations [8, 38]. More precisely, the introduction of such a minimum as a principal component in various stochastic discounting models makes quite clear the use of proactivity as main factor of leading decision making in global risk severity reduction operations [7, 35, 37].

Moreover, it is also of particular importance to consider the contribution in global risk governance operations of the maximum of a random number of positive random variables. We suppose that the positive random variables denote the severities of a random number of competing global risks. In this case, the maximum of a random number of independent positive random variables represents the worst scenario for the evolution of the corresponding global risks. Hence the combination of a result of pure probability theory with the concept of global risk governance substantially contributes to the description and solution of a problem arising in the area of global risks. It is easily adopted that the maximum of a random number of positive independent random variables can be introduced as a principal component in the formulation of stochastic models suitable for development and implementation of proactive global risk severity reduction operations. The introduction of such a maximum in stochastic modeling is mainly based on the practical interpretation on the incorporated positive independent random variables in various areas of global risk governance.

Global risk severity reduction operations are generally considered as powerful practical activities for proactive or reactive treatment of global risks. The description, investigation, and implementation of such an operation require incorporation of qualitative and quantitative concepts. It is readily understood that the qualitative concepts are elements of global risk governance and the quantitative concepts are elements of pure probability theory [48]. Since the present work concentrates on the contribution of probability theory in advancing the discipline of global risk governance it is very useful to consider significant probability results in the area of global risk severity reduction operations. Below, we present the use of a probabilistic result in implementation of a proactive global risk severity reduction operation. The severity of a global risk can be represented by a positive random variable. We suppose that a proactive risk severity reduction operation is applied to the global risk. It is easily seen that the severity of the global risk, after the application of the risk severity reduction operation, can be also represented by another positive random variable being the product of the positive random variable, denoting the severity of the global risk before the application of the risk severity reduction operation, with a random variable taking values in the standard unit interval,. Such products of random variables or equivalently such multiplicative stochastic models are called random contractions [29, 30]. From a theoretical point of view random contractions are particularly useful in investigating unimodality, stability, selfdecomposability, infinite divisibility and other significant properties of probability distributions [58]. Moreover, from a practical point of view random contractions have been used as analytical tools in economics, management, operational research, cindynics, informatics, systemics, engineering, strategic thinking, risk management and other practical disciplines. It is of some interest to mention that random contractions can be used as principal components in the formulation of stochastic models suitable in description, investigation, and implementation of global risk governance operations. Part of the present work is devoted to the establishment of sufficient conditions for embedding a stochastic discounting model in the class of random contractions which can contribute to the practical applicability of proactive treatment of various global risks.

Global risk duration is defined as the time interval of acting for the cause of a global risk. From this definition it is easily seen that global risk duration reduction belongs to the class of proactive global risk severity reduction operations. In consequence, the use of a random contraction based on the duration of a global risk can be used for describing, investigating, and implementing a global risk reduction operation. From the definition of global risk duration and the definition of global risk duration reduction arise the presence of proactivity in every global duration reduction operation. The proactive character of a global risk duration reduction operation substantially support leading decision makers, in the discipline of global risk governance, to develop combinations of such proactive operations with reactive global risk severity reduction operations in order to make feasible the main purposes of global risk treatment operations. Frequently, the development of such combinations is particularly useful in treatment of global risks with very high severity. More precisely, these combinations give priority to the implementation of global risk duration reduction operations and then such combinations make use of reactive global risk severity reduction operations. It seems to be of particular theoretical and practical importance to consider the role of global risk frequency reduction operations in real world situations [51].

Global risk frequency denotes the number of occurrences of a global risk in a given time interval. From the definition of global risk frequency it easily follows that this concept can be represented by a discrete random variable. Moreover, the probabilistic behavior of global risk frequency is described by the corresponding probability generating function [36]. In addition, from the definition of global risk frequency it is quite obvious that proactivity is a strong characteristic of global risk frequency reduction operations. In consequence, the proactive character of global risk frequency reduction operations facilitates modelers and practitioners to deeply understand the significant role of such operations in treatment of global risks. The qualitative factors of proactive global risk frequency reduction operations are incorporated in the conditions and the cause of a global risk. Furthermore, the quantitative factors of proactive global risk frequency reduction operations are covered by the results of pure probability theory [45].

An interesting interpretation of a result of probability theory in global risk frequency reduction operations is the following. We suppose that a discrete random variable denotes the frequency of a global risk and we also consider a positive random variable following the standard uniform distribution. In addition, we suppose that the discrete random variable and the standard uniform random variable are independent. It is obvious that the product of these random variables is a positive random variable. Furthermore, the integral part of the above mentioned product of random variables is a discrete random variable. If we apply a global risk frequency reduction operation to that global risk and the effect of the applied operation on the frequency of the global risk is described by the standard uniform random variable then the formulated integral part can be interpreted as the frequency of the global risk after the application of the global risk frequency reduction operation [12, 13]. In some sense the above mentioned integral part can be considered as the discrete analogue of the uniform random contraction [34]. In consequence, the integral parts can be used as principal components in the formulation of stochastic models for fundamental global risk governance operations. From the fact that the probability function and the probability generating function of an integral part are known then the probabilistic behavior of the frequency of a global risk, after the application of a global risk reduction operation, can deeply investigated by making use of these functions. Clearly the presence of unique mode at the point zero for the probability function of an integral part implies that the application of an global risk governance operation having as best scenario the avoidance of the global risk can be described, investigated, and implemented by the incorporation of an integral part [34]. It is particularly important to mention that several generalizations of an integral part can substantially facilitate the formulations of new stochastic models for developments of effective global risk governance operations. The class of a—monotone random variables can be considered as an extension of the class of integral parts [57]. From a theoretical point of view the class of a – monotone random variables have attracted the interest of several researchers [1, 64]. Consequently, it seems to be of some interest the investigation of the a – monotone random variables from the point of view of the proactive global risk governance operations.

In conclusion, this section mainly concentrates on global risk governance operations which can be developed, investigated, and implemented by incorporating fundamental results of probability theory. Moreover, this section also makes use of known global risk governance operations as motives for development and investigation of new probabilistic results. In other words, the section is devoted to reveal the strong connections between the theoretical discipline of probability theory and the practical discipline of global risk governance. It is readily recognized that the establishment of such connections is an extremely difficult but exceptionally useful activity. Moreover, the interconnections between probability theory and global risk governance can be discovered by the strong collaboration between practical researchers in the discipline of global risk governance and the theoretical researchers in the discipline of probability theory. It is of some interest to mention that the incorporation of important probabilistic results in quantitative description, stochastic investigation, and implementation of proactive treatment of global risks must constitute the most valuable component of the activity for revealing the interconnections between global risk governance and probability theory.

1.4 Information Dimensions of Global Risk Governance

Information in all its form is a valuable asset to any organization. It is the basis on which strategic decisions are made and daily tasks are performed. The value of information as an asset extends beyond its value. Information is the currency of modern society, and it is hard to think of many public or private sector services that do not depend upon it. When we use the term information, in information science, we should always keep in mind that information is what is informative for a given person. What is informative depends on the interpretative needs and skills of the individual although these are often shared with members of a discourse community.

There are various definitions of the term information, but we will mention one that is applicable to the needs of information science. Information is our perception of the outer world that is acceptable for living organisms or control machines [44]. An advantage of defining information as perception of pattern is that it provides a starting point for quantifying and valuing information, making this type of definition useful for the discipline of engineering and the discipline of economics. Definitions of information that treat it as perception of pattern broaden the concept of information by adding context. Information from this perspective has a past and a future is affected by motive and other environmental and causal factors, and itself has effects. We note three properties of information which are particular significant for the implementation of the various purposes of the present work. Due to the perception of the outer world, information improves knowledge of the recipient. Information must be acceptable to the user. There must be a sender and a recipient for the existence of information.

A question naturally arises as to whether information is of objective or subjective character. It is not difficult to conclude that the information content is objective and that its value is subjective, since it depends on the information recipient. This problem can be studied only if we are able to measure information. For that purpose, a degree of uncertainty from the recipient's viewpoint for the measure of information was proposed by Shannon. The greater the value of information the less its expectancy, in other words, the less the value of information, the greater its expectancy. To measure information, it is necessary to introduce a measuring unit called a bit. The value of one bit has information about the event that may or may not take place with the same probability. A shortcoming of such approach to information measurement is the need for a greater number of repetitions necessary to assess the occurrence or non-occurrence of the event. The consequence of that is a narrow scope of application of a probabilistic approach to the development of information theory. This approach is used in a wide range of applications in the study of communications systems. Nowadays there are various attempts trying to create other approaches to information theory. However, for the time being, the approach proposed by Shannon seems to be most constructive.

The concept of information as we use it in everyday language, in the sense knowledge communicated, plays a central role in contemporary society. The development and widespread use of computer networks since the end of the Second World War, and the emergence of information science as a discipline in the middle fifties, are evidence of this focus. Although knowledge and its communication are basic phenomena of every human society, it is the rise of information technology and its global impacts that characterize ours as an information society. It is commonplace to consider information as a basic condition for economic development together with capital, labor, and raw material but what makes information especially significant at present is its digital nature. The impact of information technology on the natural and social sciences has made this everyday notion a highly controversial concept. The contribution to communication theory due to Shannon is a landmark work, referring to the common use of information with its semantic and pragmatic dimensions, while at the same redefining the concept within an engineering framework [55].The fact that the concept of knowledge communication has been designated with the word information seems, prima facie, a linguistic happenstance.

For information science it is of course important how fundamental terms are defined and in information science, as in other fields, the question of how define information is often raised. This section constitutes an attempt to review the status of the concept of information in information science of information, with reference also to interdisciplinary trends. In scientific discourse, theoretical concepts are not true or false elements or glimpses of some element of reality. Rather, they are constructions designed to do a job the best possible way. Different conceptions of fundamental terms like information are thus more or less fruitful, depending on the theories, and in the end the practical actions, they are expected to support. The use of the word information points to a specific perspective from which the concept of knowledge communication has been defined. This perspective incorporates such characteristic as novelty. More precisely, it refers to the process of knowledge transformation, and particularly to selection and interpretation within a specific context. This discussion leads to the questions of why and when this meaning was designated with the word information.

Discussions about the concept of information in other disciplines are very important for information science because many theories and approaches in information science have their origins elsewhere. The epistemological concept of information brings into play nonhuman information processes, particularly in physics and biology. And vice versa: the psychic and sociological processes of selection and interpretation may be considered using objective parameters, leaving aside the semantic dimension, or more precisely, by considering objective or situational parameters of interpretation. This concept can be illustrated also in physical terms with regard to release mechanisms, as we suggest. An overview of the concept of information in the natural sciences as well as in the humanities and social sciences cannot hope to be comprehensive. In most cases, we can only refer to fragments of theories.

Starting with an objectivist view from the world of information theory and cybernetics, information science has turned more to the phenomena of relevance and interpretation as basic aspects of the concept of information. This change is in no way a turn to a subjectivist theory, but an appraisal of different perspectives that may determine in a particular context what is being considered as informative or a document [23]. Different concepts of information within information science reflect tensions between a subjective and an objective approach. The concept of interpretation or selection may be considered to be the bridge between two poles. It is important, to consider the different professions involved with the interpretation and selection of knowledge. The most important thing in information science is to consider information as constitutive force in society and, thus, recognize the teleological nature of global information risk governance [22].

Almost every scientific discipline today uses the concept of information within its own context and with regard to specific phenomena. The notion of information has been taken to characterize a measure of physical organization, a pattern of communication between source and receiver, a form of control and feedback, the probability of a message being transmitted over a communication channel, the content of a cognitive state, the meaning of a linguistic form, or the reduction of an uncertainty. These concepts of information are defined in various theories such as physics, thermodynamics, communication theory, cybernetics, global governance, statistical information theory, psychology, inductive logic, and so on. There seems to be no unique idea of information upon which these various concepts converge and hence no proprietary theory of information [19]. A broad philosophical debate continues as to whether the concept should address a knowledge process including, as a necessary condition, a human knower or, at the very least, an interpretative system, or whether it should exclude mental states and user related intentions and be considered as addressing an objective magnitude or property of beings [52]. Between these two positions are different kinds of mediating theories, including the quest for a unified theory of information [31]. This controversy reflects the complex history of the term. The philosophical controversy about the concept of information in the last century has its origin in cybernetics, because the concepts of communication and information were conceived at a higher level of abstraction and not reduced to the communication of human knowledge [63]. Information science is only one discipline among a network of disciplines related to communication, technology, systems and governance processes. We should try to further clarify our identity and our specific goals and to strengthen the historical continuity in the field. We should not consider a concept as information in isolation, but see it in relation to other concepts such as, for example, documents, media and governance. The concept of information may have had a positive effect as status-raising concept used by professions primary working with documents. However, this positive effect has had very unfortunate side effects in raising a level of confusion in the global information risk governance. Researchers should be explicit on how define this and other theoretical terms. There are many concepts of information, and they are embedded in more or less explicit theoretical structures. In investigating information it is easy to get lost. It is generally recognized that the most important distinction in the concepts of information is the distinction between information as an object or a thing and information as a subjective concept, information as a sign, or equivalently, as depending on the interpretation of a cognitive agent. The interpretation view removes the attention from attributes of things to the release mechanisms for whom those attributes are of importance. This may call frustration, because it is difficult and because it implies teleological principles which are foreign to positivist principles of science. It is relatively easy to count the number of words in a document or describe it in other ways. It is much more difficult to try to figure out for whom that document is of importance, and what the important questions are, that the document can answer. The questions of interpretation are also difficult because we often confuse interpretation with an individualist approach. Meaning is, however, determined in social and cultural contexts.

Finally we want to emphasize the need to explicate the foundations of knowledge claims. When we represent data in our information systems, we do that in order to support certain governance activities. We should not just regard our representations as objective, because this implies that we never fully specify the theoretical, social, and historical assumptions on which we act. Al kinds of information systems have policies and more or less explicit local or global goals. What we regard as information should finally also be a reflection of the social role of information system. As information systems becomes more global and merged implicit information is often lost. This means a challenge for information science that should open its perspective to the social and cultural impact of interpretation processes taking also into account the qualitative differences between different contexts and global goals. Building global networks is basically an interpretation process. Building a scientific network as self-reflective activity presupposes the task of clarifying some common concepts. One of them is the concept is of information.

The present work incorporates research activities in sixteen areas. The first includes the description, analysis, and applicability of the fundamental quantitative components of the extremely important concept of global information risk. The second area covers the clarification of the significant role of the goals and operations of the discipline of global risk governance in strategic thinking, decision making, and evolution of modern complex organizations. The investigation of proactivity as a structural element of global information risk governance supporting the intelligent behavior of processes and systems is the third area. The fourth area contains the contribution of stochastic models as strong analytical tools of global information risk governance. The fifth consists of the recognition of random contractions as stochastic multiplicative models particularly suitable for developing, investigating, and applying proactive information risk governance operations. The presentations and interpretations in proactive global information risk governance of six classes of random contractions constitute the sixth area [59]. The seventh area incorporates the formulation of a stochastic multiplicative model incorporating two positive random variables. The structural elements of the primary random variable of the model are a discrete random variable, a sequence of discrete random variables, and a sequence of positive random variables. The secondary random variable of the model is a random variable taking values in the standard unit interval. The eighth area consists of the establishment of sufficient conditions in order the formulated stochastic multiplicative model to be a random contraction of a positive random sum. The ninth area is the evaluation of the characteristic function of the random contraction of the positive random sum. The interpretation of the formulated random contraction of the positive random sum as a stochastic discounting model constitutes the tenth area. The eleventh area involves the evaluation of the characteristic function of the stochastic discounting model arising as an interpretation of the random contraction of the eralized random sum. The twelfth area includes the establishment of sufficient conditions for embedding the characteristic function of such a stochastic discounting model into the important class of characteristic functions corresponding to generalized unimodal probability distributions. The thirteenth area contains the interpretation of the formulated contraction of the positive random sum in global information risk control operations. More precisely, the first part of the third chapter of the present work concentrates on the applications of such a contraction in development, investigation, and implementation of global information risk severity reduction operations. The structural elements and the mathematical form of that contraction substantially facilitate the proactive treatment of the severity of global information risks. From a practical point of view the involvement of the thirteenth area can be considered as a very significant contribution of this work. The fourteenth area includes the applicability of the stochastic discounting model, arising as an interpretation of the random contraction of a positive random sum, in proactive decision making related to the occurrence of global information risks. In particular, the inclusion of the fourteenth area provides risk analysts, risk managers, and other risk experts with valuable financial knowledge for the evolution of an organization threatened by global information risks. From a practical point of view this area can be recognized as a significant factor to the strategic consideration of the proactive global information risk treatment operations. The use of fundamental concepts of probability theory in formulating new random contractions for the description, investigation, and implementation of proactive global information risk operations is the fifteenth area. The sixteenth area provides reasons supporting the global information risk governance discipline as an extremely powerful tool for strategic thinking and decision making of modern complex organizations. It seems that the formulation, investigation, and applications of stochastic models can contribute to the evolution of global information risk governance as a driving force of contemporary society. Moreover, disciplines such as cindynics, systemics, strategic management, global governance, operational research, cybernetics, chaos theory, cognitive science, international economics, psychology, sociology, geoengineering, environmental science, peace studies can be proved valuable for the advancement of global information risk governance. In particular, the conceptual framework of cindynics is readily recognized as a structural factor for creating new and significant purposes of global information risk governance. This conceptual framework contains the following eight components. A set of axioms on complex systems, a set of axioms for rationality operating in complex combination of networks, a set of axioms specific of cindynics, a concept of cindynic hyperspace associated with the net work, a concept of cindynic situation, a concept of a transformation of cindynic situations, a concept of cindynic dissonance, and a concept cindynic event. These eight basic components of the cindynic conceptual framework provide analysts, modelers, and risk managers with powerful analytical tools for formulating stochastic models describing the performance of modern complex organizations threatened by global information risks [33].

Global information risk encompasses all the challenges that result from the need of an organization to control and protect information. Since global information risk has potentially critical consequences it should be approached in much the same way as any other global risk. Consequently, strategic direction is required. In addition, strategic direction is required because how an organization addresses global information risk will reflect every changing demands and the complex dynamics of the global environment. However, global information risk is not as visible as it should be, and therefore not always treated. The pace of technological change in the information age means new global information risks can appear quickly, and may not be visible as other risks. Senior staff may wrongly assume global information risks are secondary, and of less strategic importance. Global information risks have the same characteristics as other risks, and need to be treated with the same strategic consideration. The guardianship and treatment of information in all its aspects is crucial to private and public delivery. Today, more than ever, access to electronic information is vital to an organization. Carefully assessing an organization against global information risks is the first stage in identifying where an organization is most vulnerable and defining a roadmap for implementing global governance controls and monitoring to protect the information assets of that organization.

It is quite obvious that the views of the global information risk governance community about the characteristics and treatment operations of several types of global information risks can substantially contribute to describing, analyzing, and handling real world situations incorporating a wide variety of global information risks. Some such views are the following.

If the email gets deleted, the attachment is gone for good. Users, driven by delivery pressures and performance controls, often bypass or ignore good practices needed for compliance policies. Using process-controlled, automated declaration and classification procedures for capturing both paper and electronic records, the risk of critical information not being captured into the system is significantly mitigated.

In order to avoid the risk of captured information being accidentally removed from the system, organizations must invest time in selecting the right storage, availability, and disaster recovery architectures. In a controlled environment, the system also needs to provide specific "hold" or "freeze" mechanisms which prevent normal information disposition schedules from inadvertently removing critical information, for example, when litigation is in progress.

The risk of information being deliberately removed, corrupted or damaged, is defined in legal context as "spoliation of evidence" which is the "destruction or significant alteration of evidence, or the failure to preserve property for another's use as evidence in pending or reasonably foreseeable litigation". Records are a key part of any successful legal or regulatory defense, but organizations must have the ability to locate and produce their records with the assurance that they have not been altered. In order to substantially reduce the risk of spoliation, information needs to be captured in a controlled environment where access or deletion of records is only possible through the defined and security controlled disposal process. All access to records must be monitored by making use of a detailed audit log.

For information to be relevant and useful to the business, the organization needs to ensure that not only documents and content be retained and managed securely, but the context or information used to describe them also need to be carefully managed too. Equivalently, the organization has to take decisions concerning the treatment of the risk of losing the context and metadata describing the information. This is especially true in large modern organizations where content may be captured through many different systems and sit in different repositories, but is openly available across the organization.

Information needs to be available to the right people only, for the right use and at the right time. Equivalently, an organization has to concentrate on reducing the risk of information being accessed by unauthorized persons. Lax security can not only compromise confidential or sensitive commercial information but also personal details. Modern complex organizations require sophisticated security policies to stop access to information by any unauthorized person, as well as mechanisms to prevent authorized persons taking the information outside the authorized domain.

There is very little value for information that is carefully preserved for posterity, but is not available when you need it to make a decision. Information technology systems in general are a key operational risk for the organization, posing a threat to business continuity. But whereas loss of electronic process and transaction handling could be temporarily replaced with manual processes, critical information that is locked away in a system that is unavailable, cannot be manually retrieved. Information availability should be managed within the context of an overall business continuity planning. Equivalently, the organization has to concentrate on reducing the risk of technical failures, preventing access to the information. Information availability should be managed within the context of an overall business continuity planning.

In most business environments today, information is generated, received or contained in a multitude of electronic mediums, formats, and storage devices. This explosive growth is an additional source of information risk. Being able to locate the correct information within the required timescales, be it a telephone enquiry from a customer, or weeklong regulatory audit, is critical. Organizations can employ techniques such as automated content capture, classification and federation, to ensure that all relevant information is discoverable within short timescales. More precisely, organizations have to concentrate on reducing the risk of information being lost inside the digital landfill due to lack of sufficient classification.

Format refresh is a particular issue with electronically stored information. Information that is locked into obsolete mediums or proprietary formats and systems, is worthless to an organization. So long term preservation, media refresh, and format refresh need to be considered proactively. Information strategies that include the use of format standards and audited content refresh cycles will ensure that information remains accessible for the whole period that it is being kept for. Consequently, an organization has to concentrate on reducing the risk of information becoming inaccessible due to its medium or format.

Moreover, it is generally recognized that the concept of global information risk constitutes a fundamental structural factor of modern information society. This means that the identification, measurement, and treatment of global information risk can substantially support the evolution of modern complex organizations.Incorporating results of various theoretical and practical disciplines in investigating qualitative and quantitative characteristics of the concept of global information risk can strongly support the development and implementation of operations for proactive treatment of that concept. In particular systemics, operations research, stochastic modeling, and cindynics are scientific disciplines which can strongly support research activities concentrating on the nature and proactive treatment of global information risk [17].

Risk management may be defined as the systematic process of managing the risks threatening an organization in order to accomplish its goals in a way consistent with common interest, human safety, environmental factors, and the law. It consists of the planning, organizing, directing, and controlling operations, called safety or equivalently risk management operations, undertaken with the aim of developing an efficient program that decreases the negative results of risks threatening an organization. Risk identification, risk measurement, and risk treatment constitutes the fundamental risk management operations. Moreover, risk treatment is generally recognized as the most important risk management operation. Risk control and risk finance, different but interconnected activities are the two structural methods for the treatment of risk. First, the risk manager can make use of risk control operations to modify the risk exposures in such a way as to mitigate liability and personal risks of the organization, or to make the annual risk occurrences more foreseeable. Risk avoidance, risk frequency reduction, risk severity reduction, risk duration reduction, risk separation, risk combination, and some risk transfers are the basic risk control operations. Second, the risk manager can apply risk financing operations to finance the risks that do occur. The operations in this second category incorporate those risk transfers that are not considered risk control and risk retention operations. The risk manager should very carefully consider the application of at least one risk control operation to understand if it would be proper. Unless the risk exposure is avoided, the risk manager must make use of at least one risk financing operation.

Global information risk governance adapts the generic process of global risk governance and applies it to the integrity, availability, and confidentiality of information assets and the information environment. Global information risk governance should be incorporated in all decisions, in every activity and if effectively used can be a tool for treating global information risk proactively than reactively. Effective information risk governance practices should support and contribute to the overall security, culture, operational, and business processes of organizations. The decisions underpinning global information risk governance need to be consistent with operational and strategic goals and priorities of organizations. Moreover, organizations need to determine critical factors that may support or weaken their ability to manage information securely. A systematic and logical approach to global information risk governance is needed regardless of whether global information risk is being assessed for a large project implementation, for every day operational controls and processes, or the implementation of new or revised information standards. If implemented and treated appropriately, the global information risk governance process will not only identify and keep undesirable events from affecting the performance of organizations, but also identify opportunities that may lead to gain or advantage. To sustain commitment to the process and performance of global information risk governance, the process can also be formally linked to the outputs and performance measures of organizations. Effective implementation of global information risk governance means that the responsibilities for performing tasks and monitoring risks need to be clearly defined. Global information risk governance coordination falls across all areas of organizations and all staff have some responsibility for treating global information risks in their business environments. Resourcing requirements for implementing, monitoring and reviewing global information risk strategies, should be identified as a part of planning for information security and management process [6].

The role of global information risk governance, in the first instance, is to identify the valuable information that a global organization holds, calculate the level of risk to that information, understand how such risks would affect the business, and on that basis prioritizes certain information and risks. All this should occur before the information security team is brought in to put the mitigation operations in action, in line with available resources. Information risk management follows information as it is created, distributed, stored, copied, transformed, and interacted with throughout its life cycle. Global information risk governance is not something you do once, it is a living process. The purpose of identifying, assessing, and prioritizing information and risks should not be solely defensive, focused on protecting against information loss. Rather, the role of effective global information risk governance is to understand the risk appetite of the organization and implement controls proportionate with it. The controls should be proportionate with the perceived value of the information to the organization and the need of the organization to use the information in the business. The controls and approaches used to treat global risk often include a significant cultural dimension. Major risks usually require staff awareness and appropriate behaviors in order to be controlled. For example treating the risk of fraud, or damaging external relationships or reputation is always critically dependent of staff and culture. Treating information risk is no different. Business processes can significantly reduce risk if, for example, system design takes into account information content and how the information will be used. At the same time information risk is dependent upon embedding behavior and culture within the organization. The purpose of information risk governance is to reduce the information risks threatening an organization to an acceptable level and to keep them under control in a manageable way, rather try to eliminate them entirely. Good information risk governance helps an organization get the best out its information and allows it to move forward and develop, confident that its risks are under control. As the value of information and the nature of risks evolve, managers should put in place rules, procedures, and processes to monitor and control information risks across the organization on an ongoing basis.

It is absolutely necessary the presentation of the main purposes and methods of information security in the last part of this section. The absolute necessity of such a presentation is substantial based on the general recognition of information security as an area of modern information risk management with particular theoretical and practical applicability. More precisely, it is shown that several concepts and tools of information security can contribute to the evolution of global information risk governance as a discipline facilitating strategic thinking and decision making.

Information security is important in proportion to the dependence of an organization on information technology. When the information of an organization is exposed to risk, the use of information security technology is obviously appropriate. Current information security technology, however, deals with only a small fraction of global information risk. In fact, the evidence increasingly suggests that information security technology does not reduce information risk very effectively. It is quite clear that the approach to information security from the ground up must be reconsidered if an effective deal with the problem of information risk constitutes a strategic goal of an organization. Information security is required because the technology applied to information creates risks. Broadly, information might be improperly disclosed, modified in an inappropriate way, or destroyed or lost. Failures of information security are clearly adverse events which cause losses to organizations. Consequently, information security is a risk management discipline, whose purpose is to manage the cost of information risk to an organization. Where information risk is well enough understood and at least in broad terms stable, information security starts with policies. Once an information security policy has been defined, the next task is to enforce the policy. To do this, the organization deploys a mixture of processes and technical mechanisms. The final information security task is an audit to determine the effectiveness of the measures taken to protect information against risk. It is readily understood that the job of information risk management is never done. The policy definition, protection, and audit tasks are performed, and audit tasks are performed over and over again, and the lessons learned each time through the cycle are applied during the next cycle. It is increasingly evident that information security as defined above simply is not doing the job. Information security fails because it focuses on only a small part of the problem of information risk and it does not do a very good job of protecting organizations against even that small part. Information security technology focuses primarily on risk mitigation. Information security risk analysis processes are geared toward imagining and confirming technical vulnerabilities in information systems, so that steps can be taken to mitigate the risks those vulnerabilities create. In some cases management will be asked to sign acceptance after a risk analysis. A risk acceptance will typically include either a plan for future mitigation or a justification of the economic rationale for choosing not to mitigate. Information security as a discipline is often biased, toward technological mechanisms rather than process mechanisms, in favor of logical mechanisms, and against physical mechanisms. Even within the category of risk minimization activities, information security focuses more on reducing probability of an adverse event than on reducing consequences. And where consequence reduction is implemented, it tends to focus much more strongly on quick recovery than on minimizing the severity of a loss through operations to prevent damage from spreading. Information security activities rarely include any discussions of indemnity or liability transfer.

Today, information risk governance professionals have training but often no formal information risk governance education. Information risk should be treated by professionals with specialized professional education, revocable licenses to practice, ethical obligation to implement information risk operations appropriately and keep their private information in confidence, professional obligation to control the use of potentially harmful information risk treatment operations, and professional obligation to report important public information to the proper authorities. Particularly important are the ethical to make use of only suitable information risk treatment operations and protect confidentiality of organizations treated, and the professional obligation to report information to proper authorities. The obligation of risk treatment professionals to treat appropriately, and control the use of potentially harmful information risk treatment operations, will require assessing the costs and benefits of information risk treatment operations such as liability transfer, indemnification, and retention as well as mitigation, detection and response as well as prevention, and procedural as well as technical treatments. Selection of global information risk treatment operations should be based on the welfare of the organization which will maximized by optimizing cost of global information risk to the organization rather than on minimizing probability of information risk occurrence. Needless to say, the information risk professional will be obliged to avoid the implementation of information risk treatment operations whose effectiveness is demonstrably low.

Professional training in management of information security should present a broad integrated view of treatment operations, rather than the one dimensional vulnerability and mitigation focus common today. At the simplest level, this means that information security education should include financial and legal disciplines in addition to the technical disciplines taught today. Some risk management experts have begun to describe how risk management activities can be integrated across the entire spectrum of business risks [39, 43]. Information security education should be built on this kind of comprehensive framework.

Today, almost all information security assessments use qualitative rather quantitative methods. Some risk analysis methodologies and standards already incorporate rudimentary loss and expectation methods, but these are usually limited to a low, medium, and high categorization with monetary ranges assigned to the categories. Some industries have already quantified intellectual property risk in financial terms and taken decisions to manage risk by making use of financial instrument. Risk assessment findings are essentially never shared with anyone except the organization being assessed, and possibly its external auditors. In the future, many theorists and practitioners believe that information security assessments should focus not just on identifying information risks, but also on quantifying them. Specifically, information security should be characterized in financial terms, as annualized loss expectations. Once information risks are identified and quantified the resulting data should be reported by the information risk management professionals in a way that respects their ethical obligation to protect the privacy of those they treat to the information risk equivalent of a public health service.

Today, some data on risk prevalence and severity is collected by several organizations. However, reporting to such organizations is voluntary, and only a small sample of businesses even receives the questionnaires which these bodies use to collect their summary information. In the future, collection of data on information risk needs to be more regular, formal, and comprehensive. Information risk should be studied by an independent body with the characteristics of a public health service. This service should collect from information risk governance professionals, in a way which protects the privacy of the organizations those professionals treat, data on the prevalence of losses, the causes of losses, the effects of losses, and the effectiveness of information risk treatment operations. Such a service should analyze this data and publish the results of its analyses as a way to improve the state of information risk governance practice, and to inform public policy decisions about information risk governance. Obviously, the advanced research which drives the development of new treatment operations and deeper understanding of causes of information risks will continue to be undertaken by the academic and business communities, just as advanced medical research into new drugs and the causes of disease is undertaken by academic medical schools and pharmaceutical laboratories today.

Information risk governance professionals have no training in the design of experiments to test the effectiveness of the information risk governance operations they design and no training in publishing or reviewing the results of such experiments. In the future, the effectiveness of information security technology should be more effectively evaluated by an impartial body. Security technology development and selection should be based on quantitative observational investigations of effectiveness, not on synthetic a priori assurance of vulnerability avoidance. Probabilities of exploitation must be balanced with consequences. Observed outcomes must rule, not the emotion of a good story and the fear, uncertainty and doubt that continues to be selling proposition for most security technology. Information risk governance professionals should be professionally obliged to avoid ineffective global information risk treatment operations. The effectiveness of global information risk treatment operations will change over time as the technical environment and the information risk environment in the wild evolve. Information risk governance professionals should be required to report regularly to the evaluation body on the effectiveness of the global information risk governance operations they apply to their organizations. The evaluation body should continually update its assessments of treatment effectiveness based on the information it receives, and should distribute these updates to the community of information risk governance professionals.

Any global information risk for which there is no legally required control regime will be controlled only to the extent that the cost of control can be economically justified. The economics of controlling global information risks can be distorted by competition. Global information risk tolerant organizations may gain temporary competitive advantage against global information risk adverse organizations by spending less on control, especially for global information risks with low probability of occurrence, as long as they are lucky and the global information risks do not occur.

In consequence, while it cannot be said that every global information risk should be controlled using a monetary cost and benefit analysis framework, it can be said that all global information risks should be quantified to the greatest extent possible, regardless of the anticipated control regime. It can also be said that global information risks will be poorly understood until a much better job of quantification of economic losses is implemented. Finally, it can be said that information security countermeasures will continue to be difficult to justify in voluntary control regimes until their effectiveness can be expressed as a quantifiable reduction of economic losses.

1.5 Conclusions

Structural concepts of global information risk governance and fundamental results of probability theory are incorporated in the formulation of stochastic models suitable for proactive treatment of global information risks. The above mentioned structural concepts and the fundamental results constitute the theoretical and practical contribution of the first chapter of this work. In consequence, the contribution of formulating stochastic models concentrating on the advancement of the discipline of global risk governance is easily adopted as particularly useful in theory and practice.

References

1. M.H. Alamatsaz, On characterizations of exponential and gamma distributions. Statist. Probab. Lett. **17**(4), 315–319 (1993)
2. C.T. Artikis, P.T. Artikis, Poisson random sums in modelling operations for treatment of ongoing risk occurrences, SPOUDAI. J. Econ. Bus. **55**, 32–47 (2000)
3. P.T. Artikis, C.T. Artikis, Properties and applications in risk management operations of a stochastic discounting model. J. Stat. Manag. Syst. **8**, 317–330 (2005)
4. C.T. Artikis, P.T. Artikis, Learning processes and information management operations utilizing a class of random sums. Collnet J. Scientometr. Info. Manag. **3**, 31–38 (2009)
5. C.T. Artikis, P.T. Artikis, K. Agorastos, Properties and management applications of a modified stochastic discounting model. J. Interdiscip. Math. **13**, 661–672 (2010)
6. C.T. Artikis, P.T. Artikis, G.P. Chondrocoukis, Consideration of learning processes based on optimal operations of information systems. J. Inf. Optim. Sci. **31**, 347–357 (2010)
7. P.T. Artikis, C.T. Artikis, K. Agorastos, Incorporating concepts of extreme value theory in formulating a discounting model for making optimal decisions in competing risks management. J. Interdiscip. Math. **13**, 113–124 (2010)
8. P.T. Artikis, C.T. Artikis, K. Agorastos, Stochastic discounting of cost effective replacements of systems under competing catastrophic risks. J. Inf. Optim. Sci. **32**, 109–120 (2011)
9. C.T. Artikis, Formulating a stochastic discounting model with actuarial and risk management applications, SPOUDAI. J. Econ. Bus. **62**, 7–15 (2012)
10. P.T. Artikis, Stochastic concepts facilitating strategic thinking in global risk governance. J. Stat. Manag. Syst. **17**(2), 183–194 (2014)
11. C.T. Artikis, Formulation and applications in strategic decision making of the maximum of a random number of discounted random cash flows. J. Stat. Manag. Syst. **17**(5&6), 529–538 (2014)
12. C.T. Artikis, P.T. Artikis, *Probability Distributions in Risk Management Operations*, (Springer, 2015)

13. P.T. Artikis, Discrete stochastic models and global information risk treatment operations in strategic processes. J. Discret. Math. Scie. Cryptogr. **20**(2), 477–491 (2017)
14. P.T. Artikis, Stochastic modeling of global risk severity. J. Stat. Manag. Syst. **21**(8), 1401–1405 (2018)
15. C.T. Artikis, P.T. Artikis, Formulation and investigation of a stochastic model for proactive treatment of global risk severity. Int. J. Appl. Syst. Stud. **9**(1), 26–30 (2020)
16. C.T. Artikis, P.T. Artikis, Formulation and incorporation of a stochastic model for supporting optimal decision making in global risk governance. J. Inf. Optim. Sci. **42**(5), 1065–1073 (2021)
17. U. Beck, *The Risk Society* (Sage, London, 1991)
18. J. Beirlant, J. Teugels, P. Vynckier, *Practical Analysis of Extreme Values Leuven*, (University Press Leuven, 1996)
19. R.J. Bogdan, Grounds for cognition. *How Goal- Guided Behavior Shapes the Mind*, (Lawrence Earlbaum, NJ, 1994)
20. N. Bostrom, M. Cirkovic, Introduction, in *Global Catastrophic Risks*, ed. By N. Bostrom, M. Cirkovic (Oxford University Press, Oxford, 2008), pp. 1–29
21. N. Bostrom, Existential risk prevention as global priority. Global Pol. **4**(1), 15–31 (2013)
22. S. Braman, Defining information: an approach for policymakers. Telecommun Policy **13**, 233–242 (1989)
23. M.K. Buckland, Information as thing. J. Am. Soc. Inf. Sci. **42**, 351–360 (1991)
24. A. Cedergren, H. Tehler, Studying risk governance using a design perspective. Saf. Sci. **68**, 89–98 (2014)
25. L. Ellis, A. Haldane, F. Moshirian, Systemic risk, governance and global financial stability. J. Bank. Finance **45**, 175–181 (2014)
26. P. Embrechts, C. Kluppelberg, T. Mikosch, *Modelling extremal events for insurance and finance* (Springer, Berlin, 1997)
27. P. Embrechts, S. Resnick, G. Samorodnitsky, Extreme value theory as a risk management tool. North Am. Actuarial J. **3**, 30–41 (1999)
28. N. Ford, Modeling cognitive processes in information seeking: from popper to pask. J. Am. Soc. Inform. Sci. Technol. **55**, 769–782 (2004)
29. J. Galambos, I. Simonelli, Products of random variables. *Applications to Problems of Physics and to Arithmetical Functions.* (Marcel Dekker, New York, 2004)
30. E. Hashorva, A. Pakes, Q. Tang, Asymptotics of random contractions. Insurance Math. Econ. **47**, 405–414 (2010)
31. W. Hofkirchner, The quest for a unified theory of information, in *Proceedings of the Second International Conference on the Foundations of Information Science*, (Gordon and Breach , Amsterdam, 1999)
32. N. Johnson, S. Kotz, A. Kemp, *Univariate Discrete Distributions*, 3rd edn. (Wiley, New York, 2005)
33. G. Kervern, *Latest Advances in Cindynics* (Economica, Paris, 1994)
34. N. Krishnaji, A characteristic property of the yule distribution. Sankya A **32**, 343–346 (1970)
35. A. Klinke, O. Renn, Systemic risks as challenge for policy making in risk governance, in *Forum Qualitative Sozialforschung/Forum: Qualitative Social Research*, vol. 7(1) (2006). https://doi.org/10.17169/fqs-7.1.64
36. A. Klinke, O. Renn, Risk governance: contemporary and future challenges, in *Regulating Chemical Risks* (2010), pp. 9–27
37. A. Klinke, O. Renn, O., Adaptive and integrative governance on risk and uncertainty. J. Risk Res. **15**(3), 273–292 (2012)
38. A. Klinke, Public understanding of risk and risk governance. J. Risk Res. (2020). https://doi.org/10.1080/13669877.2020.1750464
39. J. Lam, *Enterprise Risk Management* (Wiley, New York, 2003)
40. R. Lidskog, L. Soneryd, Y. Uggla, *Transboundary Risk Governance* (Earthscan Publications Ltd., London, 2010)
41. E. Lukacs, *Characteristic Functions* (Hafner, New York, 1970)

42. S.A. Lundqvist, Why firms implement risk governance–stepping beyond traditional risk management to enterprise risk management. J. Account. Public Policy **34**(5), 441–466 (2010)
43. L. Meulbroek, The promise and challenge of integrated risk management. Risk Manag. Insur. Rev. **5**, 55–66 (2002)
44. D.C. Milic, Synergy of mathematics, informatics, cybernetics and computing. Int. J. Electr. Comput. Eng. Syst. **1**, 35–37 (2010)
45. A.N. O'Connor, *Probability Distributions used in Reliability Engineering* (University of Maryland, USA, 2011)
46. G.P. Patil, S.W. Joshi, W. Sharadchandra, International statistical institute [Corporate Author], *Dictionary and Bibliography of Discrete Distributions, Food and Agriculture Organization (FAO)*, (1968)
47. O. Renn, White paper on risk governance: towards an integrative framework, in *Global Risk Governance: Concept and Practice Using the IRGC Framework, International Risk Governance Council Bookseries*, vol. 1, ed. by O. Renn, Walker (Springer, Dordrecht, 2008), pp. 3–73
48. O. Renn, A. Klinke, Theory and modelling: risk concepts and risk classification, *Risk Analysis: Opening the Process* (1998)
49. O. Renn, A. Klinke, M. Van Asselt, M, Coping with complexity, uncertainty and ambiguity in risk governance: a synthesis. Ambio **40**, 231–246 (2011)
50. O. Renn, A. Klinke, Complexity, uncertainty and ambiguity in inclusive risk governance, in *Risk and Social Theory in Environmental Management* (2012), pp. 59–76
51. O. Renn, A. Klinke, Risk perception and its impacts on risk governance. Environ. Sci. (2012). https://doi.org/10.1093/acrefore/9780199389414.013.2
52. G. Ropohl, Der informationsbegriff im Kulturstreit [The concept of information in the framework of the culturalist struggle]. Ethik und Sozialwissenschaften **1**, 1–12 (2001)
53. S. Ross, *Applied Probability Models with Optimization Applications* (Holden-Day, San Francisco, 1970)
54. S.M. Ross, Introduction to Probability Models, Elsevier (2014)
55. C. Shannon, A mathematical theory of communication. Bell Syst. Tech. J. **27**(379–423), 623–656 (1948)
56. J. Sloan, *Learning to Think Strategically* (Butterworth-Heinemann, Burlington, MA, 2006)
57. F. Steutel, Note on discrete a—unimodality. Stat. Neerl. **42**, 137–140 (1988)
58. F. Steutel, K. van Harn, *Infinite Divisibility of Probability Distributions on the Real Line* (Marcel Dekker Inc., New York, 2004)
59. J. Tait, L. Levidow, Proactive and reactive approaches to risk. Futures **24**, 219–231 (1992)
60. H. Tijms, *Stochastic Modeling and Analysis : A Computational Approach* (Wiley, New York, 1988)
61. N. Ushakov, Selected Topics in Characteristic Functions (Brill Academic Pub., Tokyo, 1999)
62. B. Wahlstrom, Models, modeling and modelers: an application to risk analysis. Eur. J. Oper. Res. **75**, 477–487 (1994)
63. N. Wiener, *Cybernetics or Communication and Control in the Animal and the Machine*, 2nd edn. (MIT Press, Cambridge, MA, 1948)
64. F. Wu, S. Dharmadhikari, Convolutions of α-unimodal discrete distributions. Stat. Neerl. **53**(2), 247–250 (1999)
65. K. Ziha, Modeling of worsening. J. Syst. Cybern. Inform. **10**(4), 11–16 (2012)

Chapter 2
Consideration in Modeling of Random Contractions

Abstract The second chapter concentrates on the theoretical and practical applicability of random contractions. It is generally recognized that random contractions constitute stochastic multiplicative models particularly suitable for describing, analyzing and solving significant problems arising in a very wide variety of scientific disciplines. The contribution of this chapter consists of introducing eight new classes of random contractions. The results of the second chapter are useful for supporting new research activities in the areas of formulating, investigating, and applying new classes of random contractions for solving significant theoretical and practical problems arising in leading scientific disciplines such as informatics, systemics, cybernetics, operational research, risk and crisis management, and cindynics. In particular the investigation of the probabilistic behavior of the formulated extended forms of random contractions creates opportunities for further applications in theory and practice of such random contractions.

Keywords Random contraction · Distribution function · Probability density function · Characteristic function · a-unimodal · v-unimodal · Maxima · Minima · Stochastic integral · Present value · Bernoulli selection

2.1 Introduction

A model is generally considered as an abstraction. It is the representation of some aspect of the world by a simplified entity which can be more easily understood and applied to clarify a real world situation. More precisely, scientists make use of models in order to describe, clarify, predict or assess aspects of the complex reality [36]. Beyond the extremely important role of models in the scientific approach, there seems to be strong connection between our mental processes and these structures that outline, interpret, and consequently direct our ideas. Moreover, models capture the nature of human understanding and can consequently be considered as a valuable element of cognitive processes [21]. The modeling process includes steps of model formulation, model validation, and model application. Models are useful for understanding the causal procedures of a real system. Such an understanding makes

P. T. Artikis and C. T. Artikis, *Random Contractions in Global Risk Governance*, Learning and Analytics in Intelligent Systems 27,
https://doi.org/10.1007/978-3-030-95691-2_2

easier to choose inflows for affecting outflows in a beneficial direction. Models are also useful for making predictions of future behavior of systems. The predictions make it viable to efficiently choose actions which give desired results. In this way models are used as instruments for optimal control plans. Models can also be particularly useful for decision makers training. There is frequently a need to incorporate many models into a consolidated structure to provide users with different views of a system. Models can also support scientists to provide answers to questions not only about the object of knowledge but also about the knowing subject. Moreover, an important objective of a model is its communicability. Without this attribute, the model might not be much of help in raising or in resolving problems. Models should also be instrumental in communicating ideas, concepts, perceptions, and solutions between the members of interested groups. The process of modeling involves the construction of models of systems, organisms, organizations, and institutions. The increased interest in modeling is not only due to the intellectual necessity to obtain a deeper insight into the process of perception of real world situations. It has increased even more since contemporary facilities are becoming more and more capable of the creative phrases of modeling. In addition, it is now spreading into all university departments, thus assuring the continuing positive feedback growth of the practice. Now efforts have even begun to bring the act of the modeler into the elementary and secondary schools. Learning is a process of modeling where a very wide variety of models are formulated for establishing an understanding of reality [34, 39].

The use of mathematics in the investigation of real world systems has become widespread in recent times. This is partly due to the use of the scientific approach to problem solving and the increasing power of computers and the computing methodology, both of which have made many more problems amenable to a mathematical solution. The steps involved in making use of mathematics in the investigation of real world systems are problem formulation, mathematical description, mathematical analysis, and interpretation of analysis to obtain a solution. The most crucial and important step is the satisfactory translation of a system from the real world into a mathematical description. Once this is done, standard methods of mathematical analysis can be used to obtain a solution to the problem. The validity of the solution depends on how well the mathematical description represents the real world. The mathematical description is called a mathematical model and the finite sequence of activities for obtaining it is called mathematical modeling. More precisely, a mathematical model of a system is a symbolic representation incorporating an abstract mathematical formulation. Such a formulation is not a mathematical model by itself. It is only by relating the mathematical formulation to a system partial description that the mathematical formulation becomes a mathematical model. A satisfactory mathematical model is a mathematical model, which is satisfactory for the purpose in the mind of the modeler. Such a mathematical model can be achieved by relating, on a one to one basis, the features of a satisfactory system partial description with the variables of a suitable mathematical formulation [3].

Mathematical modeling is an art as well as a science. The science side deals with the topics needed to execute the various steps in the mathematical modeling process. Because no two problems are the same in the real world, features such as creativity, intuition, and foresight also play a very important role in the mathematical modeling process. The science side can be appreciated in a passive learning mode, whereas, the art part can be appreciated by constructing mathematical models and learning from experience. In general, one seldom obtains a satisfactory mathematical model at the very first attempt. This is because either the system partial description is not satisfactory or the mathematical formulation is not appropriate. This implies that an iterative procedure is required, where at each stage an improved mathematical model is built and tested for adequacy. If it passes the test the model can then be said to be a satisfactory mathematical model, if not the iterative process is continued.

There is no theory of mathematical modeling, and it is difficult to envisage the evolution of a theory in the near future. Each new mathematical modeling exercise poses new challenges. This does not mean that there is no procedure for mathematical modeling. In fact there is a basic methodology, which is common to all mathematical modeling exercises. While mathematical modeling has had a central consideration in economics, operational research, and physics for several decades, it now appears not only to have achieved a similar status in such diverse fields as software, electronics, mechanics, and other domains of engineering, but also in biological and psychological sciences as well. Several reasons were advanced to explain this situation. However, the main reason may lie in the recent dramatic development of computing capabilities, particularly in the form of microprocessors, which have enabled us to perform symbolic manipulations in a way hitherto impossible. As a result, the complexity of systems modeled and of mathematical models themselves has radically increased a complexity that has severely challenged existing mathematical procedures and has numerous contributions to support mathematical modeling activities.

Mathematical models constitute a major category of models meriting distinction. The mathematical modeler maps his perceptions of the real world onto a parsimonious set of symbols. If he can do so successfully, he then obtains the ability to extend to expand his analysis to a depth that would be out of the question with verbal methods. Mathematical models may be considered as formal logical constructs with no necessarily close relationship to any aspect of reality, although modelers who stress formalism face the event of developing sterile mathematical exercises. Many theoretical and practical disciplines deal with mathematical models, but the relevance of what can be derived from them depends upon whether their basic assumptions were a reasonable representation of some aspect of reality. When the modelers no longer question their assumptions or are uninterested in the bond between the mathematical models and the real world, then there is a possibility that what started as a bold venture into the application of mathematical models to reach a greater understanding of society will become no more than a logical game or mathematical exercise [39].

A mathematical model is said to be static if it does not contain time dependent behavior. If a time dependent behavior is included in a mathematical model then the model is said dynamic. A mathematical model is said to be deterministic if it always gives the same output when a specific input applied. Finally, a mathematical model is said stochastic if the model includes a chance mechanism. From these definitions it follows that the mathematical models can be classified into four categories based on the mathematical structure of the underlying formulation. The four categories are deterministic static, deterministic dynamic, stochastic static and stochastic dynamic mathematical models.

In real life there is always uncertainty. If the uncertainty is insignificant, then it can be ignored and the system can be described by making use of a deterministic static or deterministic dynamic mathematical model. This is a process of simplification or idealization. If the uncertainty is significant then it can not be ignored and must be taken into account. In this case the system can be described by making use of a stochastic static or stochastic dynamic model. Therefore, when uncertainty is a significant feature of the system, then the system partial description must be done in a stochastic framework. As a consequence the mathematical formulations, which can be used in mathematical modeling, must allow the variables to change in an unpredictable fashion, so that they adequately represent the uncertainty of the system when the formulation becomes a model. Such formulations belong to classical probability theory and dynamic probability theory. More precisely, when the formulation is static then it belongs to classical probability theory, and when the formulation is dynamic then it belongs to dynamic probability theory.

It is quite clear that stochastic models are more realistic than deterministic models in many situations of particular practical interest. Even then, very often a much better representation is given by considering a collection or equivalently a family of random variables instead of a single random variable. Collections or equivalently families of random variables, that indexed by a parameter such as time and space, are known as stochastic processes. It is also quite clear that the mathematical analysis of stochastic dynamic models is much more difficult than the mathematical analysis of stochastic static models. Stochastic differential equations, stochastic difference equations, and stochastic integrals are very powerful analytical tools of stochastic calculus which can substantially contribute to a deep mathematical analysis of very complex stochastic dynamic models.

It has been observed that as a system, be it social, economic or physical, gets more complex, it becomes more difficult to describe in deterministic mathematical models. In fact, probability theory and stochastic models were developed in response to the failure of deterministic mathematical models to deal with complicated systems. It is generally recognized that stochastic models constitute a class of mathematical models with extremely important applications in a very wide variety of theoretical and practical disciplines [39]. For the results to be applicable stochastic models should be used in a region where they are valid. There are many implicit stochastic models used by different actors in practical processes. It is not an easy task to ensure that these stochastic models are valid representations of their corresponding real world systems. The validity requirement can to some extent be ensured by making stochastic models very transparent and thus at least to some extent separately verifiable.

The space of stochastic models for fundamental risk governance operations gives global risk analysts very important possibilities for value creation, but this expansion also generates new obligations. Risk governance practitioners must know how to make use of these models and actively decide on their careful implementation.

Stochastic modeling in the area of global risk governance is one of the topics of academic research that is in an invariably active interaction with present advances in its domain of applications. The research in this area intends at a better understanding of the stochastic evolution of systems through the formulation of suitable stochastic models, as well as at the development of effective new methodologies for fundamental risk governance operations [9, 15].

The analytical tools for effective stochastic modeling in global risk governance are definitely accessible. In order to make stochastic models a very significant section of risk governance thinking, however, we need to actively encourage their application. Otherwise, the development of theory without practical support becomes inadequate [10, 16]. Rewarding practical success, not just format rigor, would go a long way to help assure that proper attention is paid to the development of application friendly theory. The corresponding threat is that without practical justification, or at least its promise, the theory will not be taken seriously. Where global risk governance theories are allowed to breed, unchecked by the test of applicability, they tend to proliferate. In turn, the proportion of theories to practical applications grows and the gap between theory and practice widens. On the other hand, there is not a commensurate increase in what we could reasonably qualify as real knowledge about global risk governance. Some positive steps could act to enhance the development of more applicable global risk governance theories. Promotion of global risk governance theories through research publications, funds for development and other incentives for builders of such theories must be more strongly linked to applicability. The main purpose of the present work is not to intimidate global risk governance theorists. Rather, we need to counterbalance the effects of a system that rewards theorizing for the sake of theorizing. The decision sciences, including global risk governance, however, have been slow to invoke a cut and try philosophy. Ideas in global risk governance often remain page bound. That is, they are developed on paper and rarely leave it. Clearly, the decision maker in the discipline of global risk governance does not only think in terms of steel, wood, and other tangible physical objects, yet his results do affect the physical world. The key is moving stochastic models of global information risk governance operations.

Conciliation between global risk governance theorists and global risk governance practitioners can come with wide formulation and application of effective stochastic models. Such models offer the two groups the possibility to talk about. At the same time, global risk governance theorists are grounded by a strong portion of reality and global risk governance practitioners are advanced from commonplace. The final result is actual progress in the knowledge base of global risk governance community with appearing advantages incurring to world society at large [17, 27].

The stochastic character of severity, frequency, duration, and timing of a global risk constitutes the most important factor strongly supporting the formulation, investigation, and application of stochastic models in the area of global risk governance practices. During the last two decades the expansion of global risk governance was particularly impressive. Moreover, the fundamental global risk governance operations can become more elaborate by formulating, investigating, and implementing stochastic models for solving real world problems. A thorough consideration of the role of stochastic modeling activities to the evolution of global risk governance as a discipline of extremely practical importance makes quite clear a continuing significance attached to the applications of stochastic models in measurement and particularly in proactive treatment operations for global risks [6, 17].

In consequence, the general recognition of stochastic models, as very strong analytical tools for solving important real world problems in a wide variety of leading scientific disciplines, is a good reason for undertaking research activities in the area of stochastic modeling of proactive global risk governance operations. The present work mainly concentrates on the formulation, investigation, and interpretation of a class of stochastic models suitable for describing, investigating, and applying global risk governance operations considering proactivity as the most significant of their structural factors [6].

Clearly, it is well known that stochastic models such as products of positive random variables, sums of real random variables, quotients of two positive random variables, differences of two positive random variables, sums of a random number of positive independent and identically distributed random variables, minima and maxima of a random number of positive independent and identically distributed random variables, exponential transformations of positive random variables, logarithmic transformations of positive random variables, linear transformations of real random variables, integral parts of positive random variables, integral parts of products of discrete random variables taking values in the set of natural numbers with standard uniform random variables, sums and differences of two counting processes, products of continuous in time and space stochastic processes, integrals of stochastic processes, and many other stochastic models constitute very strong analytical tools for the description of the behavior and evolution of systems arising in a very wide variety of practical disciplines [33]. A significant part of the present work concentrates on investigating the ability of a class of such models in developing, investigating, and applying proactive global risk governance operations. More precisely, this work is devoted to the formulation and investigation of stochastic models with applications in global risk severity reduction operations. Each formulated model is based on the product of two independent random variables. One of the variables is positive and the other random variable takes values in the standard unit interval. From a theoretical point of view the formulated stochastic multiplicative models are interpreted as analytical tools strongly supporting the presence of proactivity global risk treatment operations. These models are called random contractions [25].

The following chapters constitute the contribution of the present work. The main purpose of the second chapter is the consideration in modeling of eight classes of random contractions by making use of several important concepts of probability theory. The main purpose of the third chapter is the investigation of an extension of one of the formulated classes of random contractions. This class is based on a generalization of the concept of sum of a random number of positive independent and identically distributed random variables. Finally, the main purpose of the fourth chapter is to make clear that such a class of random contractions is useful for developing, investigating, and implementing global risk severity reduction operations.

2.2 Defining Random Contractions

Stochastic multiplicative models incorporating two positive and independent random variables are considered as very strong analytical tools of probability theory [22]. Moreover, these models are particularly useful for the description, analysis and control of processes and systems of the real world [37]. The present section concentrates on concepts of stochastic multiplicative models incorporating a positive random variable and a random variable taking values in the interval

$$(0, 1).$$

Let

$$H$$

be a positive random variable and

$$R$$

be a random variable taking values in the interval

$$(0, 1).$$

Moreover, we suppose that the random variable

$$H$$

is independent of the random variable

$$R.$$

We consider the stochastic multiplicative model

$$V = HR.$$

It is well known that the above stochastic multiplicative model has very significant practical applications in various disciplines. The present section is devoted to the interpretations of the formulated stochastic multiplicative model in several practical disciplines.

In income distribution analysis

$$H$$

represents true income and

$$V$$

represents reported income [28].

In inventory decision making

$$H$$

represents demand for a product within a time interval and

$$V$$

represents stock of the product within the same time interval [40].

In continuous discounting of a single cash flow under random time and constant rate of interest,

$$H$$

denotes a payment to be paid at some future time and

$$V$$

denotes the present value as viewed from time zero of the payment [19].

Similarly, in continuous discounting of a continuous uniform cash flow under random duration and unit rate of interest,

$$H$$

denotes the rate of payment of a continuous uniform cash flow and

$$V$$

denotes the present value as viewed from time zero of such a continuous uniform cash flow [5].

In reliability theory,

$$H$$

may be viewed as the life length of a system which is reduced to

$$V$$

after the occurrence of some risk [18].

In risk control

$$H$$

represents the severity of risk and

$$V$$

the severity of the same risk after a risk severity reduction operation applied [13].

Moreover, in risk control

$$H$$

represents the duration of a risk and

$$V$$

the duration of the same risk after a risk duration reduction operation applied [11, 14].

The stochastic model

$$V = HR$$

also appears naturally in the context of distribution functions with nonnegative support and unique mode at the point zero [30].

The random variable

$$V$$

is a random contraction of the random variable

$$H$$

via the random variable

$$R.$$

Significant references on the theoretical contributions in the area of random contractions are provided by [22, 25, 37].

The investigation of the probabilistic behaviour of a random contraction is based on the corresponding distribution function, probability density function and characteristic function.

2.3 Distribution Function of a Random Contraction

We suppose that

$$F_H(h), \quad h > 0,$$

is the distribution function of the random variable

$$H$$

and

$$F_R(r), \quad 0 < r < 1,$$

is the distribution function of the random variable

$$R.$$

From the assumption that the random variables

$$H, R$$

are independent it is easily shown that the distribution function of the random variable

$$V = HR$$

has the form

$$F_V(v) = \int_0^1 F_H\left({}^{v}\!/_{r}\right) dF_R(r), \quad v > 0. \tag{2.3.1}$$

2.4 Probability Density Function of a Random Contraction

We suppose that

$$f_H(h), \quad h > 0$$

is the probability density function of the random variable

$$H$$

and

$$f_R(r), 0 < r < 1,$$

is the probability density function of the random variable

$$R.$$

From (2.3.1) it follows that

$$f_V(v) = \int_0^1 \frac{1}{r} f_H(v/r) f_R(r) dr, \; v > 0, \tag{2.4.1}$$

is the probability density function of the random variable

$$V = HR.$$

2.5 Characteristic Function of a Random Contraction

We consider the characteristic function

$$\varphi_H(u)$$

of the random variable

$$H,$$

the distribution function

$$F_R(r)$$

and the probability density function

$$f_R(r)$$

of the random variable

$$R.$$

From the independence of the random variables

$$H, R$$

it is easily shown that the characteristic function of the random variable

$$V = HR,$$

has the form

$$\varphi_V(u) = \int_0^1 \varphi_H(ur)dF_R(r) \tag{2.5.1}$$

or equivalently the form

$$\varphi_V(u) = \int_0^1 \varphi_H(ur)f_R(r)dr. \tag{2.5.2}$$

From (2.3.1), (2.4.1) and (2.5.1) it is readily understood that the formula

$$F_V(v) = \int_0^1 F_H(v/r)dF_R(r)$$

is equivalent to the formula

$$F_V(v) = \int_0^\infty F_R(v/h)dF_H(h), \tag{2.5.3}$$

the formula

$$f_V(v) = \int_0^1 \frac{1}{r} f_H(v/r)f_R(r)dr,$$

is equivalent to the formula

$$f_V(v) = \int_0^\infty \frac{1}{h} f_R(v/h)f_H(h)dh, \tag{2.5.4}$$

and the formula

$$\varphi_V(u) = \int_0^1 \varphi_H(ur)f_R(r)dr,$$

is equivalent to the formula

$$\varphi_V(u) = \int_0^\infty \varphi_R(uh) f_H(h) dh \tag{2.5.5}$$

where

$$\varphi_R(u)$$

is the characteristic function of the random variable

$$R.$$

From a mathematical point of view it is quite obvious that formulas (2.3.1), (2.4.1) and (2.5.1) or (2.5.2) are more suitable than formulas (2.5.3), (2.5.4) and (2.5.5) respectively for investigating the probabilistic behavior of a random contraction.

The present work mainly concentrates on the formulation, investigation and applications in informatics of random contractions of a given type of random sum. In general, an explicit analytical evaluation of the distribution function of a random sum is not possible. On the contrary, an explicit analytical evaluation of the characteristic function of a random sum is always possible. This means that the present work must consider, instead of distribution functions, characteristic functions which are the Fourier transforms of distribution functions. It is generally recognized that characteristic functions are powerful tools for the investigation of distributions. The uniqueness theorem, the convolution theorem, and the continuity theorem are very significant theorems which describe strong connections between characteristic functions and distribution functions. These theorems account for the importance of characteristic functions in probability theory [30].

The contribution of the characteristic function in investigating a stochastic model consists of providing a method for obtaining the distribution function corresponding to the stochastic model. The method is based on the Fourier inversion of the characteristic function of such a model. More precisely, the modeler can obtain the distribution function of the model by first obtaining and then inverting the corresponding characteristic function. Moreover, the use of a computational algorithm known as the Fast Fourier Transform makes the inversion of the characteristic function more manageable by greatly reducing computational time. It is generally accepted that Fourier analysis is a very powerful methodology and it could eventually revolutionalise the investigation of distribution functions corresponding to complicated stochastic models. Applications of Fourier analysis in stochastic modeling activities have been slowed by the relatively advanced nature of mathematics it entails and by the absence of computer programs that have been adapted for convenient use in various practical disciplines. As these barriers are overcome, application of Fourier analysis to solve significant problems in various practical disciplines will become more common.

The random contractions of the present work incorporate two random variables and two sequences of random variables as structural elements. Moreover, two random sums constitute the fundamental factors of these random contractions. From these facts it follows that characteristic functions and their Fourier inversions are the main analytical tools for investigating and applying in informatics of the formulated random contractions. The investigation of the probabilistic behaviour of the random contraction

$$V = HR,$$

with the random variable

$$H$$

being a function of random variables, is extremely difficult. Such an investigation is substantially facilitated by the presence of theoretical properties in the distribution function

$$F_V(v).$$

More precisely the presence of unimodality, infinite divisibility, selfdecomposability, stability and other theoretical properties in the above distribution function supports analysts, modelers and decision makers for investigating the probabilistic behaviour of the random contraction

$$V = HR.$$

In particular the unimodality and infinite divisibility of the distribution function

$$F_V(v)$$

make the random contraction

$$V = HR$$

particularly useful for the description, analysis and implementation of the most probable or extreme practical scenarios.

The present work concentrates on random contractions

$$V = HR$$

with the random variable

$$H$$

to be based on two random sums. More precisely the random variable

$$H$$

is a random sum incorporating another random sum. It is quite obvious that the corresponding characteristic function

$$\varphi_V(u),$$

will be the main analytical tool for investigating the probabilistic behavior of such random contractions. From a theoretical point of view, the present work is devoted to the establishment of sufficient conditions for the structural elements of the random sum

$$H,$$

incorporating another random sum, to be independent of the random variable

$$R.$$

2.6 Several Fundamental Classes of Random Contractions

From a theoretical point of view it is quite obvious that the number of random contractions

$$V = HR$$

which can be formulated is infinite. Since the practical applicability of a random contraction defines its importance it is very useful to mention several classes of random contractions with particular practical interest.

The following classes of random contractions are strongly supported by theoretical and applied research.

- Class of a-Unimodal Random Contractions
- Class of v-Unimodal Random Contractions
- Class of Random Contractions of Maxima of Random Numbers of Random Variables
- Class of Random Contractions of Minima of Random Numbers of Random Variables
- Class of Random Contractions of Present Value Stochastic Integrals
- Class of Random Contractions of Random Sums
- Class of Random Contractions of Present Values
- Class of Random of Random Contractions of Bernoulli Selections

The present work concentrates on some members of the sixth class of random contractions. More precisely, this work concentrates on the formulation, investigation and applications in informatics of a random sum which is based on another random sum.

2.7 Class of a-Unimodal Random Contractions

We suppose that the random variable

$$R$$

follows the beta distribution with probability density function

$$f_R(r) = ar^{a-1}, \quad a > 0, \quad 0 < r < 1. \tag{2.7.1}$$

From (2.5.2) and (2.7.1) it follows that the characteristic function of the random contraction

$$V = HR$$

has the form

$$\varphi_V(u) = a \int_0^1 \varphi_H(ur) r^{a-1} dr, \tag{2.7.2}$$

Characteristic functions of the form (2.7.2) are called a-unimodal. Similarly distribution functions, probability density functions and random variables with characteristic function of the form (2.7.2) are called a-unimodal [32]. Hence the a-unimodal random variable

$$V$$

is a random contraction of the random variable

$$H$$

via the random variable

$$R$$

following the beta distribution with probability density function given by (2.7.1). For the set of such random contractions we use the term "Class of a-Unimodal Random Contractions".

If

$$a \leq 1$$

then the probability density function

$$f_V(v) = a \int_0^1 \frac{1}{r} f_H\left(v/_r\right) r^{a-1} dr,$$

of the a-unimodal random contraction

$$V = HR$$

has a unique mode at the point 0 [31].

This means that the class of a-unimodal random contractions, with

$$a \leq 1,$$

provides modelers, analysts and decision makers with very strong analytical tools for describing and investigating an area right to the point zero as the most probable scenario of random processes.

Below we consider a particular case of the random contraction of the present section.

We suppose that the random variable

$$H$$

follows the exponential distribution with characteristic function

$$\varphi_H(u) = \frac{\mu}{\mu - iu} \tag{2.7.3}$$

and the random variable

$$R$$

follows the beta distribution with distribution function

$$F_R(r) = r^2, \quad 0 < r < 1. \tag{2.7.4}$$

From (2.7.2), (2.7.3) and (2.7.4) it follows that the characteristic function of the random contraction

$$V = HR$$

has the form

$$\varphi_V(u) = 2\int_0^1 \frac{\mu}{\mu - iur} r dr \tag{2.7.5}$$

Integrating in (2.7.5) we get that

$$\varphi_V(u) = \frac{\frac{\mu}{iu} log\left(\frac{\mu}{\mu - iu}\right) - 1}{iu/2\mu} \tag{2.7.6}$$

The characteristic function in (2.7.6) is the renewal characteristic function corresponding to the characteristic function

$$\frac{\mu}{iu} log\left(\frac{\mu}{\mu - iu}\right). \tag{2.7.7}$$

The characteristic function in (2.7.7) belongs to the product of two independently distributed random variables one of which follows the exponential distribution with parameter

$$\mu$$

and the other follows the standard uniform distribution [14].

It is very important to mention that the there are several leading publications in the area of a-unimodal random contractions [20].

Below we consider two interpretations in proactive, information risk management operations of the random contraction

$$V = HR,$$

of the random variable

$$H$$

via the random variable

$$R$$

following the beta distribution with probability function

$$f_R(r) = ar^{a-1}, \quad a > 0, \quad 0 < r < 1.$$

We suppose that the random variable

$$H$$

denotes the severity of a risk and a risk severity reduction operation is applied to the conditions and the cause of that risk. Moreover, we suppose that the impact of the risk severity reduction operation is described by the probabilistic behavior of the random variable

$$R$$

following the beta distribution with probability density function

$$f_R(r) = ar^{a-1}, \quad a > 0, \quad 0 < r < 1.$$

Hence the a-unimodal random contraction

$$V = HR$$

can be interpreted as the severity of the risk after applying the risk severity reduction operation.

Similarly, we suppose that the random variable

$$H$$

denotes the duration of a risk and a risk duration reduction operation is applied to the conditions and the cause of that risk. Moreover, we suppose that the impact of the risk duration reduction operation is described by the probabilistic behavior of the random variable

$$R$$

following the beta distribution with probability density function

$$f_R(r) = ar^{a-1}, \quad a > 0, \quad 0 < r < 1.$$

Hence the a-unimodal random contraction

$$V = HR$$

can be interpreted as the duration of the risk after applying the risk severity reduction operation.

Consequently, the class of a-unimodal random contractions is recognized as a strong analytical tool for the investigating and applying risk severity reduction operations and risk duration reduction operations.

The assumption

$$a \leq 1,$$

implies that the probability density functions of a-unimodal random contractions have a unique mode at the point 0. This assumption is particularly important in information risk management since information risk avoidance is the best and most probable scenario.

2.8 Class of v-Unimodal Random Contractions

We suppose that the random variable

$$R$$

follows the beta distribution with probability density function

$$f_R(r) = v(1-r)^{v-1}, \quad v > 0, \quad 0 < r < 1 \tag{2.8.1}$$

From (2.5.2) and (2.8.1) it follows that the characteristic function of the random contraction

$$V = HR$$

has the form

$$\varphi_V(u) = v \int_0^1 \varphi_H(ur)(1-r)^{v-1} dr. \tag{2.8.2}$$

Characteristic functions of the form (2.8.2) are called v-unimodal. Similarly distribution functions, probability density functions and random variables with characteristic function of the form (2.8.2) are called v-unimodal [23]. Hence the v-unimodal random variable

$$V$$

is a random contraction of the random variable

$$H$$

via the random variable

$$R$$

which follows the beta distribution with probability density function given by (2.8.1). For the set of such random contractions we use the term "Class of υ-Unimodal Random Contractions".

If

$$\upsilon \geq 1$$

then the probability density function

$$f_V(v) = \upsilon \int_0^1 \frac{1}{r} f_H(v/r) r^{\upsilon-1} dr,$$

of the υ-unimodal random contraction

$$V = HR$$

has a unique mode at the point 0 [31]. This means that the class of υ-unimodal random contractions, with

$$\upsilon \geq 1$$

provides modelers, analysts and decision makers with very strong analytical tools for describing and investigating an area right to the point zero as the most probable scenario of random processes.

Below we consider a particular case of the random contraction of the present section.

We suppose that the random variable

$$H$$

follows the gamma distribution with characteristic function

$$\varphi_H(u) = \left(\frac{\mu}{\mu - iu}\right)^2 \tag{2.8.3}$$

and the random variable

$$R$$

follows the beta distribution with distribution function

$$F_R(r) = 1 - (1 - r)^2, \quad 0 < r < 1. \tag{2.8.4}$$

From (2.8.2), (2.8.3) and (2.8.4) it follows that the characteristic function of the random contraction

$$V = HR$$

has the form

$$\varphi_V(u) = 2\int_0^1 \left(\frac{\mu}{\mu - iur}\right)^2 (1-r)^2 dr. \tag{2.8.5}$$

Integrating in (2.8.5) we get that

$$\varphi_V(u) = \frac{\frac{\mu}{iu} log\left(\frac{\mu}{\mu - iu}\right) - 1}{iu/2\mu} \tag{2.8.6}$$

The characteristic function in (2.8.6) coincides with the characteristic function in (2.7.6).

It is important to mention that the there are several very good publications in the area of v-unimodal random contractions [38].

Below we consider two interpretations in proactive, information risk management operations of the random contraction

$$V = HR,$$

of the random variable

$$H$$

via the random variable

$$R$$

following the beta distribution with probability function

$$f_R(r) = v(1-r)^{v-1}, \quad v > 0, \quad 0 < r < 1.$$

We suppose that the random variable

$$H$$

denotes the severity of a risk and a risk severity reduction operation is applied to the conditions and the cause of that risk. Moreover, we suppose that the impact of the risk severity reduction operation is described by the probabilistic behavior of the random variable

$$R$$

following the beta distribution with probability density function

$$f_R(r) = \upsilon(1-r)^{\upsilon-1}, \quad \upsilon > 0, \quad 0 < r < 1.$$

Hence the υ-unimodal random contraction

$$V = HR$$

can be interpreted as the severity of the risk after applying the risk severity reduction operation.

Similarly, we suppose that the random variable

$$H$$

denotes the duration of a risk and a risk duration reduction operation is applied to the conditions and the cause of that risk. Moreover, we suppose that the impact of the risk duration reduction operation is described by the probabilistic behavior of the random variable

$$R$$

following the beta distribution with probability density function

$$f_R(r) = \upsilon(1-r)^{\upsilon-1}, \quad \upsilon > 0, \quad 0 < r < 1.$$

Hence the υ-unimodal random contraction

$$V = HR$$

can be interpreted as the duration of the risk after applying the risk severity reduction operation.

Consequently, the class of υ-unimodal random contractions is recognized as a strong analytical tool for the investigating and applying risk severity reduction operations and risk duration reduction operations.

The assumption

$$\upsilon \geq 1$$

implies that the probability density functions of υ-unimodal random contractions have a unique mode at the point 0. This assumption is particularly important in information risk management since information risk avoidance is the best and most probable scenario.

2.9 Class of Random Contractions of Maxima of Random Number of Random Variables

The present section concentrates on a class of random contractions based on the concept of maximum of a random number of positive, independent and identically distributed random variables [35]. Such a class has been formulated and investigated by [10].

We suppose that

$$M$$

is a discrete random variable with values in the set

$$N = \{1, 2, ...\}.$$

and probability generating function

$$P_M(z). \tag{2.9.1}$$

We also suppose that

$$\{\Theta_m, m = 1, 2, ...\}$$

is a sequence of positive, independent random variables distributed as the random variable

$$\Theta$$

with distributions function

$$F_\Theta(\theta) \tag{2.9.2}$$

and

$$R$$

is a random variable with values in the interval

$$(0, 1)$$

and distribution function

$$F_R(r). \tag{2.9.3}$$

We consider the random variable

$$H = max\{\Theta_1, \Theta_2, \ldots, \Theta_M\}$$

and the stochastic multiplicative model

$$V = HR.$$

If

$$M, \{\Theta_m, m = 1, 2, ...\}, R$$

are independent then it can be shown that the random variable

$$H = max\{\Theta_1, \Theta_2, \ldots, \Theta_M\}$$

is independent of the random variable

$$R$$

Hence the stochastic multiplicative model

$$V = HR,$$

is a random contraction of the random variable

$$H = max\{\Theta_1, \Theta_2, \ldots, \Theta_M\}$$

via the random variable

$$R.$$

From (2.9.1) and (2.9.2) it is readily shown that the distribution function of the random variable

$$H = max\{\Theta_1, \Theta_2, \ldots, \Theta_M\}$$

is given by

$$F_H(h) = P_M(F_\Theta(h)). \tag{2.9.4}$$

Moreover from (2.3.1), (2.9.3) and (2.9.4) it follows that the distribution function of the random contraction

$$V = HR,$$

with

$$H = max\{\Theta_1, \Theta_2, \ldots, \Theta_M\}$$

is given by

$$F_V(v) = \int_0^1 P_M(F_\Theta(v/r))dF_R(r). \tag{2.9.5}$$

It is easily seen that the class of random contractions of maxima of random numbers of random variables can be useful for describing, analyzing and solving problems in various practical disciplines.

Below we consider particular cases of the random contraction of the present section.

- We suppose that the random variable

$$M$$

follows the Sibuya distribution with probability generating function

$$P_M(z) = 1 - (1 - z)^\gamma, \quad 0 < \gamma \leq 1, \tag{2.9.6}$$

the random variable

$$\Theta$$

follows the exponential distribution with distribution function

$$F_\Theta(\theta) = 1 - e^{-\mu\theta}, \theta \geq 0, \mu > 0, \tag{2.9.7}$$

and the random variable

$$R$$

follows the uniform distribution with distribution function

$$F_R(r) = r, 0 < r < 1. \tag{2.9.8}$$

From (2.9.4), (2.9.6) and (2.9.7) we get that the random variable

$$H = max\{\Theta_1, \Theta_2, \ldots, \Theta_M\}$$

is exponentially distributed with distribution function

$$F_H(h) = 1 - e^{-\mu\gamma\theta}. \tag{2.9.9}$$

Moreover, from (2.9.5), (2.9.8) and (2.9.9) we get that the distribution function of the random variable

$$V = HR$$

is given by

$$F_V(v) = \int_0^1 \left(1 - e^{\mu\gamma v/\gamma}\right) dr. \tag{2.9.10}$$

An explicit analytical evaluation of the distribution function in (2.9.10) is not possible. In this case it is advisable to evaluate the corresponding characteristic function. Since

$$\varphi_H(u) = \frac{\mu\gamma}{\mu\gamma - iu} \tag{2.9.11}$$

is the characteristic function of the distribution function in (2.9.9) then from (2.5.1), (2.9.8) and (2.9.11) it follows that the characteristic function of the random variable

$$V = HR$$

has the form

$$\varphi_V(u) = \int_0^1 \frac{\mu\gamma}{\mu\gamma - iur} dr,$$

or equivalently the form

$$\varphi_V(u) = \frac{\mu\gamma}{iur} log\left(\frac{\mu\gamma}{\mu\gamma - iu}\right).$$

The above particular case of the random contraction of the present section can be extended in the following way.

We suppose that the random variable

$$D$$

can be written in the form of the random sum

$$M = \Psi_1 + \Psi_2 + \ldots + \Psi_D$$

where the random variable

$$D$$

follows the Sibuya distribution with probability generating function

$$P_D(z) = 1 - (1 - z)^{\gamma}, 0 < \gamma \leq 1, \tag{2.9.12}$$

each random variable of the sequence

$$\{\Psi_d, d = 1, 2, \ldots\}$$

follows the Sibuya distribution with probability generating function

$$P_{\Psi}(z) = 1 - (1 - z)^{\beta}, 0 < \beta \leq 1, \tag{2.9.13}$$

the random variable

$$\Theta$$

follows the exponential distribution with distribution function

$$P_{\Theta}(\theta) = 1 - e^{\mu\theta}, \theta \geq 0, \mu > 0 \tag{2.9.14}$$

and the random variable

$$R$$

follows the uniform distribution with distribution function

$$F_R(r) = r, 0 < r < 1. \tag{2.9.15}$$

From (2.9.4), (2.9.12), (2.9.13) and (2.9.14) we get that the random variable

$$H = max\{\Theta_1, \Theta_2, \ldots, \Theta_M\}$$

is exponentially distributed with distribution function

$$F_H(h) = 1 - e^{-\mu\gamma\beta h}. \tag{2.9.16}$$

Moreover, from (2.9.5), (2.9.15) and (2.9.16) we get that the distribution function of the random variable

$$V = HR$$

is given by

$$F_V(v) = \int_0^1 \left(1 - e^{\mu\gamma\beta h/r}\right) dr. \tag{2.9.17}$$

Since an explicit analytical evaluation of the distribution function in (2.9.17) is not possible we shall evaluate the corresponding characteristic function. From the fact that

$$\frac{\mu\gamma\beta}{\mu\gamma\beta - iur} \tag{2.9.18}$$

is the characteristic function of the distribution in (2.9.17) and from (2.5.1), (2.9.15), (2.9.18) we get that the characteristic function of the random variable

$$V = HR$$

has the form

$$\varphi_V(u) = \int_0^1 \frac{\mu\gamma\beta}{\mu\gamma\beta - iur} dr$$

or equivalently the form

$$\varphi_V(u) = \frac{\mu\gamma\beta}{iu} log\left(\frac{\mu\gamma\beta}{\mu\gamma\beta - iu}\right).$$

- We suppose that the random variable

$$M$$

follows the Sibuya distribution with probability generating function

$$P_M(z) = 1 - (1 - z)^{\gamma}, 0 < \gamma \leq 1, \tag{2.9.19}$$

the random variable

$$\Theta$$

follows the beta distribution with distribution function

$$F_{\Theta}(\theta) = 1 - (1-\theta)^{\gamma}, 0 < \theta < 1, v > 0, \tag{2.9.20}$$

and the random variable

$$R$$

follows the beta distribution with distribution function

$$F_R(r) = r^{\upsilon\gamma+1}, 0 < r < 1 \tag{2.9.21}$$

From (2.9.4), (2.9.19) and (2.9.20) it follows that the random variable

$$H = max\{\Theta_1, \Theta_2, \ldots, \Theta_M\}$$

follows the beta distribution with distribution function

$$F_H(h) = 1 - (1-\theta)^{\upsilon\gamma}, 0 < h < 1 \tag{2.9.22}$$

Moreover, from (2.9.5), (2.9.21) and (2.9.22) we get that for the distribution function of the random variable

$$V = HR$$

we can successively write

$$F_V(\nu) = \int_0^1 F_H\Big(\frac{\nu}{r}\Big) dF_R(r)$$

$$F_V(\nu) = \int_0^{\nu} F_H\Big(\frac{\nu}{r}\Big) dF_R(r) + \int_{\nu}^1 F_H\Big(\frac{\nu}{r}\Big) dF_R(r),$$

$$F_V(\nu) = (\nu\gamma+1)\int_0^{\nu} r^{\nu\gamma} dr + (\nu\gamma+1)\int_{\nu}^1 \Big[1-\Big(1-\frac{\upsilon}{r}\Big)^{\nu\gamma}\Big] r^{\nu\gamma} dr. \tag{2.9.23}$$

Integrating in (2.9.23) we get that the distribution function of the random variable

$$V = HR$$

has the form

$$F_V(v) = 1 - (1 - v)^{\nu\gamma+1}, 0 < v < 1.$$

An interpretation in information systems of the random contraction

$$V = HR$$

of the random variable

$$H = max\{\Theta_1, \Theta_2, \ldots, \Theta_M\}$$

via the random variable

$$R$$

is the following.

We consider a system threatened by a risk. We suppose that the random variable

$$M$$

denotes the number of different operations of the system interrupted by the occurrence of the risk. Moreover, we suppose that the random variable

$$\Theta_M, m = 1, 2, \ldots$$

denotes the time required for the recovery of the

$$m\text{th},$$

interrupted operation of the system. Hence the random variable

$$H = max\{\Theta_1, \Theta_2, \ldots, \Theta_M\}$$

denotes the maximum recovery time, or equivalently the random variable

$$H = max\{\Theta_1, \Theta_2, \ldots, \Theta_M\}$$

denotes the recovery time of the system after the risk occurrence. If a risk duration reduction operation is applied to the conditions and the cause of the risk then the random contraction

$$V = HR,$$

of the random variable

$$H = max\{\Theta_1, \Theta_2, \ldots, \Theta_M\}$$

via the random variable

$$R,$$

can be interpreted as the recovery of the system after the risk occurrence and the proactive application of the risk duration reduction operation. It is quite obvious that the proposed interpretation for the random contraction

$$V = HR,$$

of the random variable

$$H = max\{\Theta_1, \Theta_2, \ldots, \Theta_M\}$$

via the random variable

$$R,$$

is suitable for proactive risk management practices in many kinds of systems. The mathematical structure and the above interpretation of the random variable

$$H = max\{\Theta_1, \Theta_2, \ldots, \Theta_M\}$$

are strongly related with the extremely important role of information systems in the strategic thinking and decision making of modern complex organizations. Consequently, the class of random contractions of maxima of random number of random variables can be recognized as a powerful analytical tool for undertaking research activities in various theoretical and practical areas of information systems.

2.10 Class of Random Contractions of Minima of Random Number of Random Variables

This section is devoted to a class of random contractions making use of the concept of minimum of a random number of positive, independent and identically distributed random variables [35]. The formulation and investigation of this class have been provided by [12].

We consider the discrete random variable

$$M$$

with values in the set

$$N = \{1, 2, ...\}$$

and probability generating function

$$P_M(z). \tag{2.10.1}$$

We also consider the sequence

$$\{\Theta_m, m = 1, 2, ...\},$$

of positive, independent random variables distributed as the random variable

$$\Theta$$

with distribution function

$$F_\Theta(\theta) \tag{2.10.2}$$

and the random variable

$$R$$

with values in the interval

$$(0, 1)$$

and distribution function

$$F_R(r). \tag{2.10.3}$$

Moreover, we consider the random variable

$$H = max\{\Theta_1, \Theta_2, \ldots, \Theta_M\}$$

and the stochastic multiplicative model

$$V = HR.$$

If

$$M, \{\Theta_m, m = 1, 2, ...\}, R$$

are independent then it can be proved that the random variable

$$H = max\{\Theta_1, \Theta_2, \ldots, \Theta_M\}$$

is independent of the random variable

$$R.$$

Hence the stochastic multiplicative model

$$V = HR.$$

is a random contraction of the random variable

$$H = max\{\Theta_1, \Theta_2, \ldots, \Theta_M\}$$

via the random variable

$$R.$$

From (2.10.1) and (2.10.2) it easily follows that the distribution function of the random variable

$$H = max\{\Theta_1, \Theta_2, \ldots, \Theta_M\}$$

has the form

$$F_H(h) = 1 - (P_M(1 - F_\Theta(h))). \tag{2.10.4}$$

Consequently, from (2.3.1), (2.10.3) and (2.10.4) we get that the distribution function of the random contraction

$$V = HR$$

of the random variable

$$H = max\{\Theta_1, \Theta_2, \ldots, \Theta_M\}$$

via the random variable

$$R$$

has the form

$$F_V(v) = \int_0^1 \left[1 - P_M\left(1 - F_\Theta\left(\frac{v}{r}\right)\right)\right] dF_R(r). \tag{2.10.5}$$

The applicability of the random variable

$$H = max\{\Theta_1, \Theta_2, \ldots, \Theta_M\}$$

in various practical disciplines constitutes a very good reason for considering the random contraction

$$V = HR$$

as stochastic model suitable for describing and analyzing systems and processes of the real world.

Below we consider particular a case of the random contraction of the present section.

- We suppose that the random variable

$$M$$

follows the Sibuya distribution with probability generating function

$$P_M(z) = 1 - (1 - z)^{\gamma}, 0 < \gamma \leq 1, \tag{2.10.6}$$

the random variable

$$\Theta$$

follows the beta distribution with distribution function

$$F_\Theta(\theta) = \theta^{\alpha}, 0 < \theta < 1, \alpha > 0 \tag{2.10.7}$$

and the random variable

$$R$$

follows the uniform distribution with distribution function

$$F_R(r) = r, 0 < r < 1. \tag{2.10.8}$$

From (2.10.4), (2.10.6) and (2.10.7) we get that the random variable

$$H = max\{\Theta_1, \Theta_2, \ldots, \Theta_M\}$$

follows the beta distribution with distribution function

$$F_H(h) = h^{\alpha\gamma}, 0 < h < 1, \alpha\gamma > 0. \tag{2.10.9}$$

Moreover, from (2.10.5), (2.10.8) and (2.10.9) we get that for the distribution function of the random variable

$$V = HR$$

we can successively write

$$F_V(v) = \int_0^1 F_H\left(\frac{v}{r}\right) dF_R(r)$$

$$F_V(v) = \int_0^v F_H\left(\frac{v}{r}\right) dF_R(r) + \int_v^1 F_H\left(\frac{v}{r}\right) dF_R(r),$$

$$F_V(v) = \int_0^v 1 dF_R(r) + \int_v^1 \left(\frac{v}{r}\right)^{\alpha\gamma} dr, 0 < v < 1. \quad (2.10.10)$$

Integrating in (2.10.10) we get that the distribution function of the random variable

$$V = HR$$

has the form

$$F_V(v) = \frac{v^{\alpha\gamma} - \alpha\gamma v}{1 - \alpha\gamma}, 0 < v < 1.$$

The above particular case of the random contraction of the present section can be extended in the following way.

We suppose that the random variable

$$M$$

can be written in the form of the random sum

$$M = \Psi_1 + \Psi_2 + \ldots + \Psi_D$$

where the random variable

$$D$$

follows the Sibuya distribution with probability generating function

$$P_D(z) = 1 - (1 - z)^{\gamma}, 0 < \gamma \leq 1, \quad (2.10.11)$$

each random variable of the sequence

$$\{\Psi_d, d = 1, 2, \ldots\}$$

follows the Sibuya distribution with probability generating function

$$P_\Psi(z) = 1 - (1 - z)^\beta, 0 < \beta \leq 1 \tag{2.10.12}$$

the random variable

$$\Theta$$

follows the beta distribution with distribution function

$$F_\Theta(\theta) = \theta^\alpha, 0 < \theta < 1, \alpha > 0, \tag{2.10.13}$$

and the random variable

$$R$$

follows the uniform distribution with distribution function

$$F_R(r) = r, 0 < r < 1. \tag{2.10.14}$$

From (2.10.4), (2.10.11), (2.10.12) and (2.10.13) we get that the random variable

$$H = max\{\Theta_1, \Theta_2, \ldots, \Theta_M,\}$$

follows the beta distribution with distribution function

$$F_H(h) = h^{\alpha\gamma\beta}, 0 < h < 1, \alpha\gamma\beta > 0 \tag{2.10.15}$$

Moreover, from (2.10.5), (2.10.14) and (2.10.15) we get that for the distribution function of the random variable

$$V = HR$$

we can write

$$F_V(\nu) = \int_0^\nu 1 dF_R(r) + \int_\nu^1 \left(\frac{\nu}{r}\right)^{\alpha\gamma\beta} dr, 0 < \nu < 1. \tag{2.10.16}$$

Integrating in (2.10.16) we get that the distribution function of the random variable

$$V = HR$$

has the form

$$F_V(v) = \frac{v^{\alpha\gamma\beta} - \alpha\gamma\beta v}{1 - \alpha\gamma\beta}, 0 < v < 1.$$

An interpretation in information systems of the random contraction

$$V = HR$$

of the random variable

$$H = max\{\Theta_1, \Theta_2, \ldots, \Theta_M\}$$

via the random variable

$$R$$

is the following.

At the time point 0, an organization is considering to replace its information system by a new one at some future time. At the time of replacement, the new information system is provided by a random number

$$M$$

of suppliers. We suppose that the random variable

$$\Theta_m, m = 1, 2, \ldots$$

denotes the cost of the new information system provided by the

$$m\text{th}$$

supplier. From these assumptions it follows that the random variable

$$H = max\{\Theta_1, \Theta_2, \ldots, \Theta_M\}$$

denotes the most favourable replacement cost of the information system at some future time point. If, at time 0, the organization wants to consider the most favourable replacement of the information system, from the point of view of cost, then the organization has to investigate the probabilistic behavior of the random contraction

$$V = HR$$

of the random variable

$$H = max\{\Theta_1, \Theta_2, \ldots, \Theta_M\}$$

via the random variable

$$R$$

taking values in the interval

$$(0, 1)$$

and denoting the discounting factor in continuous discounting the single cash flow

$$H = max\{\Theta_1, \Theta_2, \ldots, \Theta_M\}.$$

If the replacement of the information system is induced by the occurrence of a catastrophic risk at some future time point then the proposed interpretation of the random contraction

$$V = HR$$

of the random variable

$$H = max\{\Theta_1, \Theta_2, \ldots, \Theta_M\}$$

via the random discounting factor

$$R$$

can be of particular practial importance for decision making in the area of proactive information risk management. It is readily understood that the proposed interpretation also covers and decisions for upgrading an information system.

2.11 Class of Random Contractions of Present Value Stochastic Integrals

The present section is devoted to random contractions based on a concept of present value stochastic integral. Such a concept of stochastic integral is of some particular importance in theory of finance [24].

We suppose that

$$\{X(t), t \geq 0\}$$

is a stochastic process. The random variable

$$X(t) - X(t')$$

is called an increment over the interval

$$[\rho, t']$$

A stochastic process

$$X(t) - X(t')$$

is said to have stationary and independent increments if the distribution function of the increment

$$X(t + \tau) - X(t).$$

depends on

$$\tau,$$

but is independent of

$$t,$$

and if the increments over non-overlapping intervals are independent. Two non-overlapping intervals may have a common endpoint. A stochastic process

$$\{X(t), t \geq 0\}$$

is said to be continuous in probability at

$$t$$

If

$$\lim_{\tau \to 0} P(|X(t + \tau) - X(t)| > \varepsilon) = 0$$

for any

$$\varepsilon > 0.$$

Let

$$\{X(t), t \geq 0\}$$

be a continuous in probability stochastic process with stationary and independent increments and denote by

$$\varphi(u, \tau)$$

the characteristic function of the increment

$$X(t + \tau) - X(t).$$

It is known that

$$\varphi(u, \tau)$$

is infinite divisible and that

$$\varphi(u, \tau) = \varphi^{\tau}(u, 1)$$

[29]. For the sake of brevity we write

$$\varphi_L(u) = \varphi(u, 1)$$

where

$$\Gamma = X(t + 1) - X(t).$$

Let

$$\{X(t), t \geq 0\}$$

be a stochastic process with stationary and independent increments. We assume that

$$\mathrm{E}(\mathrm{X}(t)) = \mu t$$

and

$$V(\mathrm{X}(t)) = \sigma^2 t,$$

where

$$-\infty < \mu < \infty$$

and

$$0 < \sigma^2 < \infty.$$

Moreover, we assume that

$$\{X(t), t \geq 0\}$$

is continuous in probability and its sample paths are continuous and have left limits. We also assume that

$$\delta$$

is a positive real number. The stochastic integral

$$H = \int_0^\infty e^{-\delta t} dX(t)$$

exists in the sense of convergence in probability and is finite almost surely. In addition the distribution function

$$F_H(h)$$

of

$$H = \int_0^\infty e^{-\delta t} dX(t)$$

is continuous and

$$\varphi_H(u) = exp\left\{\int_0^\infty log\varphi_\Gamma\left(ue^{-\delta t}\right)dt\right\}$$

is its characteristic function [24]. It is easily demonstrated that the characteristic function

$$\varphi_H(u)$$

can be written in the form

$$\varphi_H(u) = \exp\left\{\frac{1}{\delta}\int_0^u \frac{\log\varphi_\Gamma(y)}{y}dy\right\} \tag{2.11.1}$$

Furthermore, we suppose that the increments

$$X(t) - X(t'), t' < t$$

of the stochastic process

$$\{X(t), t \geq 0\},$$

in the stochastic integral

$$H = \int_0^\infty e^{-\delta t} dX(t),$$

are positive and this integral is independent of the random variable

$$R$$

with values in the interval

$$(0, 1)$$

and distribution function

$$F_R(r). \tag{2.11.2}$$

Hence the stochastic multiplicative model

$$V = HR$$

is a random contraction of the stochastic integral

$$H = \int_0^\infty e^{-\delta t} dX(t)$$

via the random variable

$$R.$$

From (2.5.1), (2.11.1) and (2.11.2) it follows that the characteristic function of the random contraction

$$V = HR,$$

of the stochastic integral

$$H = \int_0^\infty e^{-\delta t} dX(t)$$

via the random variable

$$R,$$

has the form

$$\varphi_V(u) = \int_0^1 \exp\left\{\frac{1}{\delta}\int_0^{ur} \frac{\log \varphi_\Gamma(y)}{y} dy\right\} dF_R(r). \tag{2.11.3}$$

It is obvious than an explicit analytical evaluation of the characteristic function

$$\varphi_H(u) = \exp\left\{\frac{1}{\delta}\int_0^{u} \frac{\log \varphi_\Gamma(y)}{y} dy\right\}$$

of the stochastic integral

$$H = \int_0^\infty e^{-\delta t} dX(t)$$

is very difficult. Hence an explicit analytical evaluation of the characteristic function

$$\varphi_V(u) = \int_0^1 \exp\left\{\frac{1}{\delta}\int_0^{ur} \frac{\log \varphi_\Gamma(y)}{y} dy\right\} dF_R(r),$$

of the random contraction of the stochastic integral

$$H = \int_0^\infty e^{-\delta t} dX(t)$$

via the random variable

$$R$$

is extremely difficult. Moreover from the fact that an explicit analytical evaluation of the distribution function corresponding to the random contraction

$$V = HR,$$

of the stochastic integral

$$H = \int_0^\infty e^{-\delta t} dX(t)$$

via the random variable

$$R,$$

is not possible, it follows that the characteristic function

$$\varphi_V(u) = \int_0^1 \exp\left\{ \frac{1}{\delta} \int_0^{ur} \frac{\log \varphi_\Gamma(y)}{y} dy \right\} dF_R(r)$$

remains the most reliable tool for investigating the probabilistic behavior of the random contraction

$$V = HR.$$

It is readily understood that the use of the characteristic function

$$\varphi_V(u) = \int_0^1 \exp\left\{ \frac{1}{\delta} \int_0^{ur} \frac{\log \varphi_\Gamma(y)}{y} dy \right\} dF_R(r),$$

for investigating the probabilistic behavior of the random contraction

$$V = HR$$

requires to overcome the extreme difficulty of the explicit analytical evaluation of the above characteristic function. It seems that the methodology of transformations of characteristic functions can substantially contribute to explicit analytical evaluation of particular cases of the characteristic function

$$\varphi_V(u) = \int_0^1 \exp\left\{\frac{1}{\delta}\int_0^{ur} \frac{\log \varphi_\Gamma(y)}{y} dy\right\} dF_R(r).$$

During the last three decades there has been an increasing interest in transformations of characteristic functions, more precisely in operations which transform a given characteristic function into a new characteristic function. Naturally one investigates the properties of the transformed characteristic function. Many characteristic functions of theoretical and practical interest arise as transformations of other characteristic functions. Some celebrated examples are given by the normal, Laplace, exponential, gamma, geometric, uniform characteristic functions [2]. Two types of transformations of characteristic functions are considered in the literature. The first type consists of transformations which are the result of mixtures of characteristic functions. The second type consists of transformations which are not the result of mixtures of characteristic functions.

The characteristic function

$$\varphi_H(u) = \exp\left\{\frac{1}{\delta}\int_0^{u} \frac{\log \varphi_\Gamma(y)}{y} dy\right\}$$

of the stochastic integral

$$H = \int_0^\infty e^{-\delta t} dX(t)$$

is a transformation of the characteristic function

$$\varphi_\Gamma(u).$$

This transformation is of the second type.

Moreover, the characteristic function

$$\varphi_V(u) = \int_0^1 \exp\left\{\frac{1}{\delta}\int_0^{ur} \frac{\log \varphi_\Gamma(y)}{y} dy\right\} dF_R(r)$$

of the random contraction

$$V = HR$$

of the stochastic integral

$$H = \int_0^\infty e^{-\delta t} dX(t),$$

via the random variable

$$R,$$

is a transformation of the characteristic function

$$\varphi_H(u) = \exp\left\{\frac{1}{\delta}\int_0^u \frac{\log \varphi_\Gamma(y)}{y} dy\right\}.$$

This transformation is of the first type.

The methodology of transformations of characteristic functions can contribute to explicit analytical evaluations of particular cases of the characteristic function

$$\varphi_V(u) = \int_0^1 \exp\left\{\frac{1}{\delta}\int_0^{ur} \frac{\log \varphi_\Gamma(y)}{y} dy\right\} dF_R(r)$$

by formulating and solving integral equations based on transformations of the first type of the characteristic function

$$\varphi_H(u) = \exp\left\{\frac{1}{\delta}\int_0^u \frac{\log \varphi_\Gamma(y)}{y} dy\right\}$$

Below we consider a particular case of the characteristic function of the random contraction

$$V = HR$$

of the stochastic integral

$$H = \int_0^\infty e^{-\delta t} dX(t) \tag{2.11.4}$$

via the random variable

$$R.$$

We suppose that the characteristic function of the random variable

$$\Gamma = X(t+1) - X(t)$$

has the form

$$\varphi_{\Gamma}(u) = exp\left\{\theta\left[\frac{\mu}{\mu - iu} - 1\right]\right\}, \tag{2.11.5}$$

where

$$\vartheta > \delta,$$

and the random variable

$$R$$

follows the uniform distribution with distribution function

$$F_R(r), 0 < r < 1. \tag{2.11.6}$$

From (2.11.1) and (2.11.5) it readily follows that the characteristic function of the stochastic integral in (2.11.4) has the form

$$\varphi_H(u) = \left(\frac{\mu}{\mu - iu}\right)^{\frac{\vartheta}{\delta}}. \tag{2.11.7}$$

Moreover, from (2.11.3) and (2.11.6) it follows that the characteristic function of the random contraction of the stochastic integral in (2.11.4) via the standard uniform random variable has the form ,

$$\varphi_V(u) = \int_0^1 \left(\frac{\mu}{\mu - iur}\right)^{\frac{\vartheta}{\delta}} dr$$

or equivalently the form

$$\varphi_V(u) = \frac{\left(\frac{\mu}{\mu - iu}\right)^{\frac{(\vartheta-\delta)}{\delta}} - 1}{i(\vartheta - \delta)u/_{\delta\mu}}$$

The above characteristic function is the renewal characteristic function corresponding to the characteristic function

$$\left(\frac{\mu}{\mu - iu}\right)^{\frac{(\vartheta-\delta)}{\delta}}$$

of the gamma distribution with parameters

$$\mu$$

and

$$\frac{(\vartheta - \delta)}{\delta}$$

An interpretation in information systems of the random contraction

$$V = HR$$

of the stochastic integral

$$H = \int_0^\infty e^{-\delta t} dX(t)$$

via the random variable

$$\mathrm{H}$$

is the following.

The theory of finance is concerned with the determination of the value of the firm as an ongoing concern, the identification and analysis of factors with direct and indirect influence on this value, and with the evaluation of investment opportunities. The economic value of the firm as a going concern is the present value of income that the firm will generate in the future. We assume that the income of the firm is given by the stochastic process

$$\{X(t), t \geq 0\},$$

More precisely, the random variable

$$X(t)$$

represents the income generated by the firm in the time interval

$$[0, t].$$

Since the corporate firm has an indefinite life, its economic value can be approximated by the stochastic integral

$$H = \int_0^\infty e^{-\delta t} dX(t)$$

where

$$\delta$$

is the force of interest [24]. Moreover, random contraction

$$V = HR,$$

of the stochastic integral

$$H = \int_0^\infty e^{-\delta t} dX(t)$$

via the random variable

$$R$$

can be used for the approximation of the economic value of the firm under the assumption that the firm has experienced a major risk occurrence during its lifetime. From the fact that the information system of a modern complex organization has the main characteristic of a corporate firm it can be said that the proposed interpretation of the random contraction

$$V = HR$$

of the stochastic integral

$$H = \int_0^\infty e^{-\delta t} dX(t)$$

is of some importance in investigating the evolution of such a system.

2.12 Class of Random Contractions of Random Sums

The present section concentrates on a class of random contractions making use of the concept of positive random sum, or more precisely making use of the concept of sum of a random number of positive, independent and identically distributed random variables. This class has been formulated and investigated by [4, 7, 8].

We suppose that

$$M$$

is a discrete random variable with values in the set

$$\boldsymbol{N}_0 = \{0, 1, 2, ...\}$$

and probability generating function

$$P_M(z), \tag{2.12.1}$$

We also suppose that

$$\{\Theta_m, m = 1, 2, ...\},$$

is a sequence of positive, independent random variables distributed as the random variable

$$\Theta$$

with characteristic function

$$\varphi_\Theta(u) \tag{2.12.2}$$

and we consider the random variable

$$R$$

with values in the set

$$(0, 1)$$

and distribution function

$$F_R(r), \tag{2.12.3}$$

We consider the random variable

$$H = \Theta_1 + \Theta_2 + \cdots + \Theta_M$$

and the stochastic multiplicative model

$$V = HR.$$

If

$$M, \{\Theta_m, m = 1, 2, ...\}, R,$$

are independent then it can be shown that the random variable

$$H = \Theta_1 + \Theta_2 + \cdots + \Theta_M$$

is a random sum which is independent of the random variable

$$R.$$

Hence the stochastic multiplicative model

$$V = HR$$

is a random contraction of the random sum

$$H = \Theta_1 + \Theta_2 + \cdots + \Theta_M$$

via the random variable

$$R.$$

From (2.12.1) and (2.12.2) it is readily shown that the characteristic function of the random sum

$$H = \Theta_1 + \Theta_2 + \cdots + \Theta_M$$

is given by

$$\varphi_H(u) = P_M(\varphi_\Theta(u)) \tag{2.12.4}$$

Moreover from (2.5.1), (2.12.3) and (2.12.4) it readily follows that the characteristic function of the random contraction

$$V = HR,$$

with

$$H = \Theta_1 + \Theta_2 + \cdots + \Theta_M,$$

is given by

$$\varphi_V(u) = \int_0^1 P_M(\varphi_\Theta(ur))dF_R(r) \tag{2.12.5}$$

The role of the above characteristic function, in investigating the probabilistic behaviour of the random contraction of the random sum

$$H = \Theta_1 + \Theta_2 + \cdots + \Theta_M$$

via the random variable

$$R,$$

is very important since an explicit analytical evaluation of the corresponding distribution function

$$F_V(v)$$

is, in general, very difficult.

It is generally recognized that the random sum

$$H = \Theta_1 + \Theta_2 + \cdots + \Theta_M$$

substantially contributes to the description, investigation and solution of problems arising in economics, management, operations research, logistics, informatics, biology, engineering and other significant practical disciplines [33]. Such contributions support the applicability in various disciplines of the random contraction

$$V = HR$$

of the random sum

$$H = \Theta_1 + \Theta_2 + \cdots + \Theta_M$$

via the random variable

$$R.$$

The important role of the characteristic function

$$\varphi_V(u) = \int_0^1 P_M(\varphi_\Theta(ur))dF_R(r)$$

in investigating the probabilistic behaviour of the random contraction

$$V = HR$$

where

$$H = \Theta_1 + \Theta_2 + \cdots + \Theta_M$$

makes necessary the consideration of some particular cases of the above characteristic function.

- We suppose that the random variable

$$M$$

follows the Bernoulli distribution with probability generating function

$$P_M(z) = pz + q, 0 < p < 1, q = 1 - p,$$

the random variable

$$\Theta$$

follows the uniform distribution with characteristic function

$$\varphi_\Theta(u) = \frac{e^{iu} - 1}{iu}$$

and the random variable

$$R$$

follows the beta distribution with distribution function

$$F_R(r) = r^2, 0 < r < 1$$

Under the above assumptions we get that the characteristic function of the random contraction

$$V = HR$$

has the form

$$\varphi_V(u) = p\frac{2\left(e^{iu} - iu - 1\right)}{i^2u^2} + q.$$

It is easily seen that the above characteristic function belongs to a Bernoulli random sum of random variables following the beta distribution with parameters 1 and 2.

- We also suppose that the random variable

$$M$$

follows the binomial distribution with probability generating function

$$P_M(z) = (pz + q)^n, 0 < p < 1, q = 1 - p$$

the random variable

$$\Theta$$

follows the exponential distribution with characteristic function

$$\varphi_\Theta(u) = \frac{\mu}{\mu - iu}, \mu > 0,$$

and the random variable

$$R$$

follows the uniform distribution with distribution function

$$F_R(r) = r, 0 < r < 1$$

Under the above assumptions we get that the characteristic function of the random contraction

$$V = HR$$

has the form

$$\varphi_V(u) = \int_0^1 \left(p\frac{\mu}{\mu - iur} + q\right)^n dr$$

or equivalently the form

$$\varphi_V(u) = \sum_{\kappa=0}^{n} \binom{n}{\kappa} p^\kappa q^{n-\kappa} \int_0^1 \left(p\frac{\mu}{\mu - iur} + q\right)^\kappa dr$$

It is readily shown that the above characteristic function can be written in the form

$$\varphi_V(u) = q^n + npq^{n-1}\frac{\mu}{iu}\log\left(\frac{\mu}{\mu - iu}\right) + \sum_{\kappa=2}^{n}\binom{n}{\kappa}p^{\kappa}q^{n-\kappa}\frac{\left(\frac{\mu}{\mu-iu}\right)^{\kappa} - 1}{\frac{i(\kappa-1)}{\mu}}$$

- Moreover, we suppose that the random variable

$$M$$

follows the geometric distribution with probability generating function

$$P_M(z) = \frac{pz}{1-qz}, 0 < p < 1, q = 1 - p,$$

the random variable

$$\Theta$$

follows the exponential distribution with characteristic function

$$\varphi_\Theta(u) = \frac{\mu}{\mu - iu}$$

and the random variable

$$R$$

follows the uniform distribution with distribution function

$$F_R(r) = r, 0 < r < 1$$

Under the above assumptions we get that the characteristic function of the random contraction

$$V = HR$$

has the form

$$\varphi_V(u) = \int_0^1 \frac{\lambda}{\lambda - iur}dr$$

or equivalently the form

$$\varphi_V(u) = \frac{\lambda}{iu} \log\left(\frac{\mu}{\mu - iu}\right)$$

where

$$\lambda = p\mu.$$

The above particular case of the characteristic function

$$\varphi_V(u) = \int_0^1 P_M(\phi_\Theta(ur))dF_R(r),$$

corresponding to the random contraction

$$V = HR$$

of the random sum

$$H = \Theta_1 + \Theta_2 + \cdots + \Theta_M$$

via the random variable

$$R,$$

incorporates three of the most important distributions of probability theory. This fact constitutes a very good reason for concentrating on the investigation of the practical applicability of the random contraction

$$V = HR$$

of the random sum

$$H = \Theta_1 + \Theta_2 + \cdots + \Theta_M$$

via the random variable

$$H,$$

where the random variable

$$M$$

follows the geometric distribution, the random variable

$$\Theta$$

follows the exponential distribution and the random variable

$$\mathrm{H}$$

follows the standard uniform distribution.

- Finally, we suppose that the random

$$M$$

follows the Laplace distribution with probability generating function

$$P_M(z) = \left(\frac{pz}{1 - qz}\right)^{n+1},$$

where

$$0 < p < 1,$$

$$q = 1 - p$$

and

$$n$$

positive integer, the random variable

$$\Theta$$

follows the exponential distribution with characteristic function

$$\varphi_\Theta(u) = \frac{\mu}{\mu - iu}, \mu > 0$$

and the random variable

$$R$$

follows the beta distribution with distribution function

$$F_R(r) = r^n, 0 < r < 1.$$

Under the above assumptions we get that the characteristic function of the random contraction

$$V = HR$$

has the form

$$\varphi_V(u) = n\int_0^1 \left(\frac{p\mu}{p\mu - iur}\right)^{n+1} r^{n-1}dr \tag{2.12.6}$$

We shall evaluate the above integral by making use of mathematical induction. If we set

$$n = 1$$

in (2.12.6) we get that

$$\varphi_V(u) = \int_0^1 \left(\frac{p\mu}{p\mu - iur}\right)^2 dr. \tag{2.12.7}$$

From (2.12.7) it is readily shown that

$$\frac{p\mu}{p\mu - iu} = \int_0^1 \left(\frac{p\mu}{p\mu - iur}\right)^2 dr \tag{2.12.8}$$

We have to prove that

$$\left(\frac{p\mu}{p\mu - iu}\right)^n = n\int_0^1 \left(\frac{p\mu}{p\mu - iur}\right)^{n+1} r^{n-1}dr \tag{2.12.9}$$

From (2.12.8) it follows that (2.12.9) is valid for

$$n = 1.$$

We suppose that (2.12.9) is valid for

$$n = \kappa.$$

Hence

$$\left(\frac{p\mu}{p\mu - iu}\right)^{\kappa} = \kappa \int_{0}^{1} \left(\frac{p\mu}{p\mu - iur}\right)^{\kappa+1} r^{k-1} dr \tag{2.12.10}$$

and we shall prove that (2.12.9) is valid for

$$n = \kappa + 1,$$

or equivalently we shall prove that

$$\left(\frac{p\mu}{p\mu - iu}\right)^{\kappa+1} = (\kappa + 1) \int_{0}^{1} \left(\frac{p\mu}{p\mu - iur}\right)^{\kappa+2} r^{k} dr \tag{2.12.11}$$

It is easily shown that if we differentiate (2.12.10) we get (2.12.11). Hence the characteristic function of the random contraction

$$V = HR$$

has the form

$$\varphi_V(u) = \left(\frac{p\mu}{p\mu - iu}\right)^{n}$$

In conclusion, it can be said that the class of random contractions of random sums constitutes a set of very strong analytical tools for the description, analysis and control of systems arising in various practical disciplines. The present work concentrates on random contractions

$$V = HR,$$

where the random variable

$$H$$

is a random sum incorporating another random sum. More precisely, this work is devoted to the formulation, investigation and application in different disciplines of informatics of random contractions of this type.

It is well known that a random sum based on another random sum constitutes a stochastic model with significant theoretical and practical applications [26] This fact is a very reason for undertaking research activities in the area of random contractions of such random sums. In particular, the concentration of research activities on the applicability of random contractions of this kind in various areas of informatics can be particularly useful for the description, analysis and solution of problems arising in various disciplines.

An interpretation in information systems of the random contraction

$$V = HR$$

of the random sum

$$H = \Theta_1 + \Theta_2 + \cdots + \Theta_M$$

via the random variable

$$R$$

is the following.

We consider a system threatened by a risk. We suppose that the random variable

$$M$$

denotes the number of different damages due to an occurrence of the risk. Moreover, we suppose that the random variable

$$\Theta_m, m = 1, 2, ...$$

denotes the severity of the

$$m\text{th}$$

damage. Hence the random sum

$$H = \Theta_1 + \Theta_2 + \cdots + \Theta_M$$

denotes the severity of all damages due to an occurrence of the risk, or equivalently the random sum

$$H = \Theta_1 + \Theta_2 + \cdots + \Theta_M$$

denotes the total severity of an occurrence of the risk. If a risk severity reduction operation is applied to the conditions and the cause of the risk then the random contraction

$$V = HR,$$

of the random sum

$$H = \Theta_1 + \Theta_2 + \cdots + \Theta_M$$

via the random variable

$$R$$

can be interpreted as the total severity of an occurrence of the risk after applying a risk severity reduction operation to that risk.

It is readily understood that the proposed interpretation is useful for proactive risk management practices in a wide variety of systems. From the fact that proactivity is generally recognized as a very important structural element for treating information risks, it can be said that the proposed interpretation constitutes a fundamental factor substantially supporting processes and purposes of information systems of modern complex organizations.

The above interpretation in information systems of the random contraction

$$V = HR,$$

of the random sum

$$H = \Theta_1 + \Theta_2 + \cdots + \Theta_M$$

via the random variable

$$R$$

can be more useful in the following way. We apply a risk frequency reduction operation to that risk. According to that risk frequency reduction operation an occurrence of the risk is retained with probability

$$p$$

and deleted with probability

$$q = 1 - p$$

The retention-deletion of a risk occurrence is independent of the retention-deletion of any other risk occurrence. If the random variable

$$G$$

denotes the number of risk occurrences required to get the first retained risk occurrence then the random variable

$$G$$

follows the geometric distribution with probability generating function

$$P_G(z) = \frac{pz}{1 - qz}.$$

We consider the sequence of positive, independent, and identically distributed random variables

$$\{I_g, g = 1, 2, ...\}$$

with

$$I_g, g = 1, 2, ...$$

denoting the time between the

$$(g - 1)\text{th}$$

and

$$g\text{th}$$

risk occurrences. If

$$G$$

and

$$\{I_g, g = 1, 2, ...\}$$

are independent then the random sum

$$\Sigma = I_1 + I_2 + \cdots + I_G$$

denotes the time required to get the first retained risk occurrence with total severity

$$H = \Theta_1 + \Theta_2 + \cdots + \Theta_M.$$

Consequently, the proposed interpretation in information systems of the random contraction

$$V = HR,$$

of the random sum

$$H = \Theta_1 + \Theta_2 + \cdots + \Theta_M$$

via the random variable

$$R,$$

is substantially improved by incorporating the above strongly proactive risk frequency reduction operation.

2.13 Class of Random Contractions of Present Values

The present section is devoted to a class of random contractions based to the concept of continuous discounting of a single random cash flow under random time and constant force of interest. This class has been formulated and investigated by [1].

We suppose that

$$H$$

is a positive random variable with characteristic function

$$\varphi_H(u) \tag{2.13.1}$$

and

$$R$$

is a random variable with values in the interval

$$(0, 1).$$

We also suppose that the random variable

$$R$$

can be written in the form

$$R = e^{-\delta T},$$

where

$$\mathrm{T}$$

is a positive random variable with distribution function

$$F_T(t) \tag{2.13.2}$$

and

$$\delta$$

is a positive integer. From (2.13.2) it easily follows that the distribution function

$$F_R(r)$$

of the random variable

$$R = e^{-\delta T}$$

has the form

$$F_R(r) = 1 - F_T\left(-\frac{1}{\delta}\log r\right), 0 < r < 1 \tag{2.13.3}$$

If the random variable

$$H$$

is independent of the random variable

$$T$$

then the random variable

$$H$$

is also independent of the random variable

$$R = e^{-\delta T}.$$

Hence the stochastic multiplicative model

$$V = HR$$

or equivalently the stochastic multiplicative model

$$V = He^{-\delta T}.$$

is a random contraction of the random variable

$$H$$

via the random variable

$$R = e^{-\delta T}.$$

From (2.13.1), (2.13.2) and (2.13.3) it follows that the characteristic function of the random contraction

$$V = He^{-\delta T}.$$

has the form

$$\varphi_V(u) = \int_0^1 \varphi_H(ur) dF_R(r)$$

or equivalently the form

$$\varphi_V(u) = \int_0^1 \varphi_H(ur) d\left[1 - F_T\left(-\frac{1}{\delta}\log r\right)\right]. \tag{2.13.4}$$

Below we consider a particular case of the random contraction of the present section.

We suppose that the random variable

$$T$$

follows the exponential distribution with distribution function

$$F_T(t) = 1 - e^{-\mu t}, t > 0, \mu > 0. \tag{2.13.5}$$

From (2.13.3) and (2.13.5) it easily follows that the distribution function of the random variable

$$R = e^{-\delta T}$$

has the form

$$F_R(r) = r^a, 0 < r < 1 \tag{2.13.6}$$

where

$$a = {}^{\mu}/_{\delta}.$$

Moreover, we suppose that the random variable

$$H$$

is arbitrarily distributed with characteristic function

$$\varphi_H(u). \tag{2.13.7}$$

From (2.13.4), (2.13.6) and (2.13.7) it follows that the characteristic function of the random contraction

$$V = He^{-\delta T}$$

has the form

$$\varphi_V(u) = a \int_0^1 \varphi_H(ur) r^{a-1} dr$$

In this case the random contraction

$$V = He^{-\delta T}$$

is called α-unimodal.

Below we consider an interpretation in proactive risk treatment of the random contraction of the present section.

We suppose that the random variable

$$T$$

denotes the time of the occurrence of a risk. Moreover, we suppose that the random variable

$$H$$

denotes the severity of that risk occurrence. If the positive real number

$$\delta$$

denotes the force of interest in continuous compounding then the random contraction

$$V = He^{-\delta T}$$

denotes the present value, as viewed from the point zero, of the severity

$$H$$

of the risk occurring at the time point

$$T.$$

It is readily understood that the incorporation of one or two of the random variables

$$max\{\Theta_1, \Theta_2, \ldots, \Theta_M\}$$

$$min\{\Theta_1, \Theta_2, \ldots, \Theta_M\}$$

and

$$\Theta_1 + \Theta_2 + \cdots + \Theta_M$$

in the mathematical form of the stochastic multiplicative model

$$V = He^{-\delta T}$$

constitutes a very good reason for undertaking research activities concentrating on the representation of such a model as a random contraction.

2.14 Class of Random Contractions of Bernoulli Selections

The present section concentrates on a class of random contractions based on the concept of Bernoulli selection. Such a class has been formulated and investigated by [17].

We consider the discrete random variable

$$I$$

following the Bernoulli distribution with probability generating function

$$P_I(z) = 1 - \rho + \rho z, 0 < \rho < 1. \tag{2.14.1}$$

We also consider the positive random variable

$$B$$

with characteristic function

$$\varphi_B(u) \tag{2.14.2}$$

the positive random variable

$$\Xi$$

with characteristic function

$$\varphi_{\Xi}(u), \tag{2.14.3}$$

and the random variable

$$R$$

with values in the interval

$$(0, 1)$$

and distribution function

$$F_R(r) \tag{2.14.4}$$

Moreover, we consider the random variable

$$H = \begin{cases} B \ , \ I = 0 \\ \\ \Xi \ , \ I = 1 \end{cases}$$

and the stochastic multiplicative model

$$V = HR.$$

If

$$I,$$

$$B,$$

$$\Xi$$

and

$$R$$

are independent then it can be proved that the random variable

$$H = \begin{cases} B \ , I = 0 \\ \Xi \ , I = 1 \end{cases}$$

is independent of the random variable

$$R.$$

Hence the stochastic multiplicative model

$$V = HR.$$

is a random contraction of the random variable

$$H = \begin{cases} B \ , I = 0 \\ \Xi \ , I = 1 \end{cases}$$

via the random variable

$$R.$$

From the assumption that

$$I,$$

$$B,$$

$$\Xi$$

and

$$R$$

are independent it follows that

$$I,$$

$$B$$

and

$$\Xi$$

are independent. Hence (2.14.1), (2.14.2) and (2.14.3) imply that the characteristic function of the random variable

$$H = \begin{bmatrix} B \ , \ I = 0 \\ \\ \Xi \ , \ I = 1 \end{bmatrix}$$

has the form

$$\varphi_H(u) = (1 - \rho)\varphi_{\mathrm{B}}(u) + \rho\varphi_{\Xi}(u) \tag{2.14.5}$$

Consequently, from (2.5.1), (2.14.4) and (2.14.5) we get that the characteristic function of the random contraction

$$V = HR$$

of the random variable

$$H = \begin{bmatrix} B \ , \ I = 0 \\ \\ \Xi \ , \ I = 1 \end{bmatrix}$$

via the random variable

$$R$$

has the form

$$\varphi_H(u) = \int_0^1 [(1 - \rho)\varphi_B(ur) + \rho\varphi_{\Xi}(ur)]dF_R(r)$$

or equivalently the form

$$\varphi_H(u) = (1 - \rho)\int_0^1 \varphi_B(ur)dF_R(r) + \rho\int_0^1 \varphi_{\Xi}(ur)dF_R(r) \tag{2.14.6}$$

From the fact that the random contraction

$$V = HR$$

of the present section has the random variables

$$I,$$

$$B,$$

$$\Xi$$

and

$$R$$

as structural elements and the random variable

$$H = \begin{cases} B \;, \; I = 0 \\ \\ \Xi \;, \; I = 1 \end{cases}$$

as principal component it seems suitable to combine the results of the present section with the results of the previous five sections. More precisely, the replacement of the structural element

$$B$$

or the structural element

$$\Xi$$

in the principal component

$$H = \begin{cases} B \;, \; I = 0 \\ \\ \Xi \;, \; I = 1 \end{cases}$$

by one of the random variables

$$\max\{\Theta_1, \Theta_2, ..., \Theta_M\},$$

$$\min\{\Theta_1, \Theta_2, ..., \Theta_M\},$$

$$\int_0^\infty e^{-\delta t} dX(t)$$

and

$$\Theta_1 + \Theta_2 + \cdots + \Theta_M$$

substantially contributes to the formulation of random contractions with significant theoretical and practical applicability. In addition, the replacement of the random variable

$$H$$

in the stochastic discounting model

$$He^{-\delta T}$$

by one of the random variables

$$\max\{\Theta_1, \Theta_2, ..., \Theta_M\},$$

$$\min\{\Theta_1, \Theta_2, ..., \Theta_M\}$$

and

$$\Theta_1 + \Theta_2 + \cdots + \Theta_M$$

strongly supports the formulation of stochastic discounting models being random contractions with useful theoretical and practical applications.

A significant part of the present work is devoted to the formulation, investigation and application in global risk governance of random contractions of Bernoulli selections incorporating an extension of the random variable

$$\Theta_1 + \Theta_2 + \cdots + \Theta_M.$$

Below we consider a particular case of the random contraction of the resent section.

We suppose that the random variable

$$B$$

follows the exponential distribution with characteristic function

$$\varphi_B(u) = \frac{\mu}{\mu - iu} \tag{2.14.7}$$

the random variable

$$\Xi$$

follows the gamma distribution with characteristic function

$$\varphi_{\Xi}(u) = \left(\frac{\mu}{\mu - iu}\right)^2 \tag{2.14.8}$$

and the random variable

$$R$$

follows the uniform distribution with distribution function

$$F_R(r) = r, 0 < r < 1 \tag{2.14.9}$$

From (2.14.6), (2.14.7), (2.14.8) and (2.14.9) it follows that the characteristic function of the random contraction

$$V = HR$$

of the random variable

$$H = \begin{bmatrix} B \text{ , } I = 0 \\ \\ \Xi \text{ , } I = 1 \end{bmatrix}$$

via the uniformly distributed random variable

$$R$$

has the form

$$\varphi_V(u) = (1 - \rho)\frac{\mu}{iu}\log\left(\frac{\mu}{\mu - iu}\right) + \rho\frac{\mu}{\mu - iu}.$$

Below we consider an interpretation in proactive risk treatment of the random contraction

$$V = HR$$

of the random variable

$$H = \begin{bmatrix} B \text{ , } I = 0 \\ \\ \Xi \text{ , } I = 1 \end{bmatrix}$$

via the random variable

$$R.$$

We suppose that an organization will make and implement a decision at a given future time point. More precisely, the organization will start an activity with probability

$$\rho$$

or another similar activity with probability

$$1 - \rho.$$

The activity selected by the organization with probability

$$\rho$$

is threatened by a risk having severity

$$\Xi.$$

Moreover, the activity selected by the organization with probability

$$1 - \rho.$$

is threatened by a risk having severity

$$B.$$

Hence the random variable

$$H = \begin{cases} B \ , \ I = 0 \\ \\ \Xi \ , \ I = 1 \end{cases}$$

denotes the severity of risk undertaken by the organization at the given future time point. In consequence the random contraction

$$V = HR$$

can be interpreted as the severity of the risk undertaken by the organization after applying a severity reduction operation to that risk.

2.15 Conclusions

The present chapter concentrates on the theoretical and practical applicability of random contractions, which are generally recognized as very strong analytical tools of probability theory. More precisely, it is shown that random contractions constitute stochastic multiplicative models particularly suitable for describing, analyzing and solving significant problems arising in a very wide variety of scientific disciplines. The main contribution of this chapter consists of introducing eight classes of random contractions by incorporating two beta distributed random variables, a maximum of a random number of positive independent and identically distributed random variables, a minimum of a random number of positive independent and identically distributed random variables, a present value stochastic integral, a sum of a random number of positive independent and identically distributed random variables, a present value of a single cash flow and a Bernoulli selection. Conclusively, it can be said that the results of the first chapter of the present work are useful for supporting new research activities in the areas of formulating, investigating, and applying extended forms of random contractions for solving significant theoretical and practical problems arising in leading scientific disciplines such as informatics, systemics, cybernetics, operational research, risk and crisis management, and cindynics. In particular the investigation of the probabilistic behavior of the formulated extended forms of random contractions, or equivalently the establishment of theoretical properties for the corresponding probability distributions, creates opportunities for further applications in theory and practice of such random contractions.

References

1. T. Artikis, D. Jerwood, On the unimodality of present value distributions. Eur. J. Oper. Res. **52**, 107–112 (1991)
2. T. Artikis, A. Voudouri, A stochastic integral arising in discounting continuous cash flows and certain transformed characteristic functions. Appl. Math. Lett. **13**, 87–90 (2000)
3. P.T. Artikis, C.T. Artikis, Stochastic models in fundamental risk management operations. Int. Rev. Econ. **51**, 207–219 (2004)
4. C.T. Artikis, P.T. Artikis, Poisson random sums in modelling operations for treatment of ongoing risk occurrences. SPOUDAI: J. Econ. Bus. **55**, 32–47 (2005)
5. P.T. Artikis, C.T. Artikis, Properties and applications in risk management operations of a stochastic discounting model. J. Stat. Manag. Syst. **8**, 317–330 (2005)
6. P.T. Artikis, C.T. Artikis, C.E. Foundas, P. Hatzopoulos, Discrete renewal and selfdecomposable distributions in modeling information risk management operations. J. Stat. Manag. Syst. **9**, 73–85 (2006)
7. C.T. Artikis, P.T. Artikis, C.E. Fountas, Risk management operations described by a stochastic discounting model incorporating a random sum of cash flows and a random maximum of recovery times. J. Stat. Manag. Syst. **10**, 439–450 (2007)
8. C.T. Artikis, P.T. Artikis, Incorporating a random number of independent competing risks in discounting a continuous uniform cash flow with rate of payment being a random sum. J. Interdiscip. Math. **10**, 487–495 (2007)

9. P.T. Artikis, C.T. Artikis, Random sums of integral part models in computer systems operations. J. Discrete Math. Sci. Cryptogr. **11**, 209–217 (2008)
10. P.T. Artikis, C.T. Artikis, J. Moshakis, Discounted maximum of a random number of random cash flows in optimal decision making. J. Inf. Optim. Sci. **29**, 1193–1201 (2008)
11. C.T. Artikis, P.T. Artikis, Stochastic modeling of risk duration reduction operations. J. Stat. Manag. Syst. **12**, 131–139 (2009)
12. P.T. Artikis, C.T. Artikis, Discounted minimum of a random number of random variables in replacement of computer systems. J. Inf. Optim. Sci. **30**, 887–897 (2009)
13. C.T. Artikis, P.T. Artikis, C. Kara-Zaitri, Stochastic modeling of risk severity reduction operations. J. Stat. Manag. Syst. **12**, 641–650 (2009)
14. P.T. Artikis, C.T. Artikis, Stochastic models for control programs of organizations. Kybernetes **39**, 570–577 (2010)
15. P.T. Artikis, Stochastic concepts facilitating strategic thinking in global risk governance. J. Stat. Manag. Syst. **17**(2), 183–194 (2014)
16. C.T. Artikis, Formulation and applications in strategic decision making of the maximum of a random number of discounted random cash flows. J. Stat. Manag. Syst. **17**(5&6), 529–538 (2014)
17. C.T. Artikis, Random contractions in continuous discounting of Bernoulli selecting processes arising in systemics. Int. J. Appl. Syst. Stud. **6**(4), 294–301 (2016)
18. I. Bazovsky, *Reliability Theory and Practice* (Dover Publications, 2004)
19. A. De Schepper, M. Goovaerts, J. Dhane, R. Kaas, D. Vyncke, Bounds for present value functions with stochastic interest rates and stochastic volatility. Insur. Math. Econom. **31**, 87–103 (2002)
20. S. Dharmadhikari, K. Jogdeo, *Unimodality, Convexity, and Applications* (Academic Press Inc., New York, 1988)
21. N. Ford, Modeling cognitive processes in information seeking: from popper to pask. J. Am. Soc. Inform. Sci. Technol. **55**, 769–782 (2004)
22. J. Galambos, I. Simonelli, *Products of Random Variables. Applications to Problems of Physics and to Arithmetical Functions* (Marcel Dekker, New York, 2004)
23. M. Gomes, D. Pestana, The use of fractional calculus in probability. Port. Math. **37**, 259–271 (1978)
24. I. Harrison, Ruin problems with compounding assets. Stoch. Process. Appl. **5**, 67–79 (1977)
25. E. Hashorva, A. Pakes, Q. Tang, Asymptotics of random contractions. Insur. Math. Econom. **47**, 405–414 (2010)
26. N. Johnson, S. Kotz, A. Kemp, *Univariate Discrete Distributions*, 3rd edn. (Wiley, New York, 2005)
27. F. Kloman, Rethinking risk management. Geneva Papers Risk Insur. **17**, 299–313 (1992)
28. N. Krishnaji, Characterization of the Pareto distribution through a model of underreported incomes. Econometrica **38**, 251–255 (1970)
29. E. Lukacs, A characterization of stable processes. J. Appl. Probab. **6**, 409–418 (1969)
30. E. Lukacs, *Characteristic Functions* (Hafner, New York, 1970)
31. P. Medgyessy, On a new class of unimodal infinitely divisible distribution functions and related topics. Stud. Sci. Math. Hung. **2**, 441–446 (1967)
32. R. Olshen, S. Savage, A generalized unimodality. J. Appl. Probab. **7**, 21–34 (1970)
33. M. Pinsky, S. Karlin, *An Introduction to Stochastic Modeling*, 4th edn. (Elsevier Inc., Oxford, UK, 2011)
34. J. Rasmussen, Skills, rules, knowledge: signals, signs, and symbols. IEEE Trans. Syst. Man Cybern. **13**, 257–267 (1983)
35. D.E.A. Sanders, The modeling of extreme events. Br. Actuar. J. **11**, 519–557 (2005)
36. K.S. Shrader-Frechette, *Risk Analysis and Scientific Method* (D. Reidel Publishing Company, Dordrecht, 1985)
37. F. Steutel, K. van Harn, *Infinite Divisibility of Probability Distributions on the Real Line* (Marcel Dekker Inc., New York, 2004)

38. A. Voudouri, *Continuous Univariate Distibutions Arising in Finance*, Ph.D. thesis, Postgraduate School of Studies in Mathematical Sciences, University of Bradford (1988)
39. B. Wahlstrom, Models, modeling and modelers: an application to risk analysis. Eur. J. Oper. Res. **75**, 477–487 (1994)
40. D. Waters, *Inventory Control and Management* (Wiley, New York, 2007)

Chapter 3
Formulations and Investigations of Random Contractions

Abstract The contribution of the third chapter consists of the implementation of several theoretical and practical purposes related to the formulation of a stochastic multiplicative model, the representation of the formulated stochastic multiplicative model as a random contraction, the explicit analytical evaluation of the characteristic function corresponding to such a random contraction, the interpretation of the formulated stochastic multiplicative model and the related random contraction as a stochastic discounting model, and the explicit analytical evaluation of the characteristic function corresponding to such a stochastic discounting model. The chapter also establishes an extension of the formulated stochastic multiplicative model by incorporating a positive random variable and a Bernoulli random variable and the representation of the formulated stochastic model as a random contraction. The explicit analytical evaluation of the characteristic function and the establishment of a characterization of such an extension constitute new results of the third chapter. The chapter also includes the representation of the extension of the formulated stochastic multiplicative model and the related random contraction as a stochastic discounting model. The explicit analytical evaluation of the characteristic function is also provided.

Keywords Random contraction · Stochastic multiplicative model · Stochastic discounting · Characteristic function

3.1 Introduction

The contribution of stochastic models to description, analysis, and solution of problems arising in various practical disciplines is generally recognized as very important. Such models provide designers of systems with extremely useful probabilistic information for investigating and controlling the performance of a wide variety of systems with significant applications in many practical disciplines. During the last five decades, the disciplines of probability and statistics have made quite clear that random sums are very powerful stochastic models. More precisely, these disciplines have heavily concentrated on the suitability of random sums to support fundamental

P. T. Artikis and C. T. Artikis, *Random Contractions in Global Risk Governance*, Learning and Analytics in Intelligent Systems 27,
https://doi.org/10.1007/978-3-030-95691-2_3

operations of systems [5]. It is very well known that systemic thinking in management, economics, reliability, insurance, engineering, cindynics, and many other practical disciplines is facilitated by the formulation, investigation, and application of random sums [6].

Random sums are very powerful to quite a wide range of fundamental branches of probability theory. It is generally accepted that theory of risk and insurance, reliability, stochastic theory of service systems, stochastic processes, mixtures of probability distributions, stochastic integrals, random sampling and statistical inference are the most celebrated branches of probability theory and statistics which make use of random sums. Moreover, random sums have very important applications in a variety of significant practical disciplines. Stochastic modeling activities arising in physics, chemistry, informatics, biology, medicine, neurophysiology, psychology, economics, logistics, general management, engineering, and other disciplines make substantial use of random sums. Such practical disciplines take advantage of theoretical results, numerical methods, and computer simulation techniques related to random sums, and provide analysts, modelers and decision makers with more detailed and more realistic criteria upon which to base analysis, predictions and decision making under conditions of uncertainty. It is of some particular importance to note that Poisson random sums are the most significant cases of random sums from a theoretical and practical point of view.

The applicability of a random sum is mainly based on the existence of an explicit form for the corresponding characteristic function and the use of the Fast Fourier Transform. These two factors are of great significance for the applicability of random sums with very complicated mathematical structure. The consideration and practical applicability in global risk governance of some particular stochastic multiplicative models, incorporating random sums, constitute the main research purposes of the present work. It is easily understood that the mathematical structure of the concept of random sum can substantially facilitate various extensions of such a concept. More precisely, the recognition of a discrete random variable taking values in the set of nonnegative integers and a sequence of arbitrary random variables as structural elements of the mathematical form of a random sum make quite clear the extremely significant role of these elements in formulating various extensions of random sums. This recognition implies the extensions of random sums, suitable for investigating the performance and evolution of systems arising in various practical disciplines, can only be developed by proposing transformations for the constituent elements of such sums. It is obvious that the possibility of proposing complicated transformations for the constituent elements of random sums directly leads to developments of extensions of random sums which are particularly useful for the description, analysis, and control of systems of significant applicability for a wide variety of practical disciplines [1]. In constructing random sums for the description and analysis of random phenomena it is very useful to start with some basic random variables taken as having known properties and to combine these random variables systematically in order to obtain random variables of particular applicability, the combination being in physically meaningful terms. The second chapter of this work introduces such combinations for developing random sums supporting applications of random contractions in global risk governance. Considering the results of the present chapter from a practical point of view then the following conclusion can be taken. Random sums constitute a class of stochastic models incorporating extremely valuable information being particularly suitable for formulating random contractions which are fundamental analytical tools in global risk governance operations.

The present chapter concentrates on the implementation of the following purposes. The first purpose is the formulation of a stochastic multiplicative model by making use of an extension of the concept of sum of a random number of positive random variables and a random variable taking values in the standard unit interval. The second purpose is the establishment of sufficient conditions for representing the formulated stochastic multiplicative model as a random contraction. The third purpose is the evaluation of the characteristic function of such a random contraction. The fourth purpose is the interpretation of the formulated stochastic multiplicative model and the related random contraction as a stochastic discounting model. The fifth purpose is the evaluation of the characteristic function of such a stochastic discounting model. The sixth purpose is an extension of the formulated stochastic multiplicative model by incorporating a positive random variable and a Bernoulli random variable. The seventh purpose is the establishment of sufficient conditions for representing the extension of the formulated stochastic model as a random contraction. The evaluation of the characteristic function and the establishment of a characterization of such an extension constitute the eighth and the ninth purpose respectively of the present chapter. The tenth purpose is the representation of the extension of the formulated stochastic multiplicative model and the related random contraction as a stochastic discounting model. Moreover, the evaluation of the characteristic function and the establishment of a characterization of such a stochastic discounting model constitute the eleventh and the twelfth purpose respectively of the present chapter. The implementations of the above mentioned twelve purposes are completely devoted to make quite clear the role of the concept of sum of a random number of positive, independent, and identically distributed random variables in substantially supporting the theoretical and practical applicability of the concept of random contraction. More precisely, the present chapter mainly seeks to clarify that the structural elements and the mathematical form of a sum of a random number of positive, independent, and identically distributed random variables constitute the fundamental factors for formulating, investigating, and applying random contractions of such a sum. It is readily understood that the complexity of these factors directly affects the theoretical investigation and the practical applicability of these random contractions. The concept of sum of a random number of positive random variables which is driven by a sum of a random number of discrete random variables, and the concept of Bernoulli selecting process are generally recognized as important analytical tools of probability theory with interesting applications in a wide variety of practical disciplines. A significant part of the present chapter makes use of these two analytical tools for formulating and investigating random contractions. The structural elements and the mathematical form of such random contractions provide analysts, modelers, and decision makers with valuable information for investigating the performance and evolution of modern complex organizations.

3.2 Formulation of a Stochastic Multiplicative Model

The present section formulates a stochastic multiplicative model by using a discrete random variable, a sequence of discrete random variables, a sequence of positive random variables, and random variable with values in the standard unit interval.

We consider the discrete random variable

$$N$$

with values in the set

$$\mathbf{N} = \{1, 2, ..\},$$

the sequence

$$\{L_n, n = 1, 2, \dots\}$$

of discrete random variables with values in the set

$$N_0,$$

and the discrete random variable

$$S = L_1 + L_2 + \cdots + L_N.$$

We also consider the sequence

$$\{X_s, s = 1, 2, \dots\}$$

of positive random variables, the positive random variable

$$C = X_1 + X_2 + \cdots + X_S,$$

and the positive random variable

$$W$$

with values in the interval

$$(0, 1).$$

Moreover, we consider the stochastic multiplicative model

$$Y = CW.$$

The practical and theoretical applications of the above stochastic multiplicative model are mainly based on the discrete random variable

$$S = L_1 + L_2 + \cdots + L_N$$

and the positive random variable

$$C = X_1 + X_2 + \cdots + X_S.$$

3.3 Representation as a Random Contraction of a Stochastic Multiplicative Model

The establishment of sufficient conditions for representing the stochastic multiplicative model

$$Y = CW$$

as a random contraction of the random variable

$$C = X_1 + X_2 + \cdots + X_S$$

via the random variable

$$W$$

is a purpose of the present section.

Theorem 3.3.1 *Let*

$$N$$

be a discrete random variable with values in the set

$$\mathbf{N}_0$$

and probability generating function

$$P_N(\mathrm{z}),$$

and let

$$\{L_n, n = 1, 2, \ldots\}$$

be a sequence of discrete and independent random variables with values in the set

$$N_0.$$

We suppose that the random variables of the sequences

$$\{L_n, n = 1, 2, \ldots\}$$

are equally distributed with the random variable

$$L$$

having probability generating function

$$P_L(z)$$

and we consider the random variable

$$S = L_1 + L_2 + \cdots + L_N.$$

Let

$$\{X_s, s = 1, 2, \ldots\}$$

be a sequence of positive and independent random variables. We suppose that the random variables of the sequence

$$\{X_s, s = 1, 2, \ldots\}$$

are equally distributed with the random variable

$$X$$

having characteristic function

$$\varphi_X(u).$$

We set

$$C = X_1 + X_2 + \cdots + X_S$$

and let

$$W$$

be a random variable with characteristic function

$$\varphi_W(\pi)$$

and values in the interval

$$(0, 1).$$

If

$$N,$$

$$\{L_n, n = 1, 2, \ldots\},$$

$$\{X_s, s = 1, 2, \ldots\},$$

and

$$W$$

are independent then the stochastic multiplicative model

$$Y = CW$$

is a random contraction of the random variable

$$C = X_1 + X_2 + \cdots + X_S$$

via the random variable

$$W.$$

Proof From the assumption that

$$N,$$

$$\{L_n, n = 1, 2, \ldots\},$$

$$\{X_s, s = 1, 2, \ldots\},$$

and

$$W$$

are independent it follows that the random variable

$$N$$

and the sequence

$$\{L_n, n = 1, 2, \dots\}$$

are independent. Hence the discrete random variable

$$S = L_1 + L_2 + \cdots + L_N$$

is a random sum with probability generating function

$$P_S(z) = P_N(P_L(N)). \tag{3.3.1}$$

Moreover, if

$$\varphi_L(u)$$

is the characteristic function of the random variable

$$L$$

then the characteristic function of the random sum

$$S = L_1 + L_2 + \cdots + L_N$$

is given by

$$\varphi_S(u) = P_N(\varphi_L(u)).$$

The evaluation of the characteristic function

$$\varphi_C(u),$$

of the random variable

$$C = X_1 + X_2 + \cdots + X_S$$

requires the establishment of the independence of the sequence

$$\{X_s, s = 1, 2, \dots\}$$

of positive, independent, and identically distributed random variables and the random sum

$$S = L_1 + L_2 + \cdots + L_N.$$

This means the establishment of the independence of the random variables

$$X_1, X_2, ..., X_s, S.$$

This establishment requires the establishment of the relationship

$$\varphi_{X_1,...,X_s,S}(u_1, u_2, ..., u_s, \xi) = \varphi_{X_1}(u_1)...\varphi_{X_s}(u_s)\varphi_S(\xi), \tag{3.3.2}$$

where

$$\varphi_{X_1,...,X_s,S}(u_1, u_2, ..., u_s, \xi)$$

is the joint characteristic function of the random variables

$$X_1, X_2, ..., X_s, S.$$

We have

$$\varphi_{X_1,...,X_s,S}(u_1, u_2, ..., u_s, \xi) = E\left(e^{iu_1X_1+\cdots+iu_sX_s+i\xi S}\right)$$

or equivalently

$$\varphi_{X_1,...,X_s,S}(u_1, u_2, ..., u_s, \xi) = \sum_{n=0}^{\infty} E\left(e^{iu_1X_1+\cdots iu_sX_s+i\xi S}|N = n\right)P(N = n) \tag{3.3.3}$$

From (3.3.3) it follows that

$$\varphi_{X_1,...,X_s,S}(u_1, u_2, ..., u_s, \xi) = \sum_{n=0}^{\infty} E\left(e^{iu_1X_1+\cdots iu_sX_s+i\xi L_1+\cdot+i\xi L_n}|N = n\right)P(N = n). \tag{3.3.4}$$

Since

$$N,$$

$$\{L_n, n = 1, 2, \ldots\},$$

and

$$\{X_s, s = 1, 2, \ldots\}$$

are independent then (3.3.4) can be written in the form

$$\varphi_{X_1,\ldots,X_S,S}(u_1, u_2, \ldots, u_s, \xi) = \sum_{n=0}^{\infty} E\left(e^{iu_1X_1+\cdots iu_sX_s+i\xi L_1+\cdots+i\xi L_n}\right)P(N = n). \tag{3.3.5}$$

Since the sequences

$$\{L_n, n = 1, 2, \ldots\},$$

$$\{X_s, s = 1, 2, \ldots\},$$

are independent then the random variables

$$X_1, \ldots, X_s, L_1, \ldots, L_n,$$

are independent and the random variables

$$e^{iu_1X_1}, \ldots, e^{iu_sX_s}, e^{i\xi L_1}, \ldots, e^{i\xi L_n}$$

are also independent. Then (3.3.5) can be written in the form

$$\varphi_{X_1,\ldots,X_s,S}(u_1, u_2, \ldots, u_s, \xi) = \sum_{n=0}^{\infty} E\left(e^{iu_1X_1}\right)\ldots E\left(e^{iu_SX_s}\right)E\left(e^{i\xi L_1}\right)\ldots E\left(e^{i\xi L_n}\right)P(N = n). \tag{3.3.6}$$

From (3.3.6) it follows that

$$\varphi_{X_1,\ldots,X_s,S}(u_1, u_2, \ldots, u_s, \xi) = E\left(e^{iu_1X_1}\right)\ldots E\left(e^{iu_sX_s}\right)\sum_{n=0}^{\infty} E\left(e^{i\xi L_1}\right)\ldots E\left(e^{i\xi L_n}\right)P(N = n). \tag{3.3.7}$$

Since

$$\varphi_L(\xi) = E\left(e^{i\xi L_\kappa}\right), \kappa = 1, 2, \ldots, n$$

then (3.3.7) can be written in the form

$$\varphi_{X_1,\ldots,X_s,S}(u_1, u_2, \ldots, u_s, \xi) = \varphi_{X_1}(u_1)\ldots\varphi_{X_s}(u_s)\sum_{n=0}^{\infty} \varphi_L^n(\xi)P(N = n). \tag{3.3.8}$$

Since

$$\varphi_S(\xi) = \sum_{n=0}^{\infty} \varphi_L^n(\xi) P(N = n)$$

or equivalently

$$\varphi_S(\xi) = P_N(\varphi_L(\xi)),$$

is the characteristic function of the random sum

$$S = L_1 + L_2 + \cdots + L_N$$

then (3.3.8) can be written in the form

$$\varphi_{X_1,\dots,X_s,S}(u_1, u_2, \dots, u_s, \xi) = \varphi_{X_1}(u_1)\dots\varphi_{X_s}(u_s)\varphi_S(\xi). \tag{3.3.9}$$

From (3.3.9) it follows that (3.3.2) is valid. Hence the sequence

$$\{X_s, s = 1, 2, \dots\}$$

is independent of the random sum

$$S = L_1 + L_2 + \cdots + L_N.$$

Consequently, for the characteristic function

$$\varphi_C(u)$$

of the random sum

$$C = X_1 + X_2 + \cdots + X_S,$$

called random sum in the present work, we have

$$\varphi_C(u) = E\left(e^{iuC}\right),$$

or equivalently

$$\varphi_C(u) = \sum_{s=0}^{\infty} E\left(e^{iuC} | S = s\right) P(S = s). \tag{3.3.10}$$

From (3.3.10) it follows that

$$\varphi_C(u) = \sum_{s=0}^{\infty} E\big(e^{iuX_1}...e^{iuX_s}|S = s\big)P(S = s). \tag{3.3.11}$$

Since it has been proved that the random sum

$$S = L_1 + L_2 + \cdots + L_N$$

is independent of the sequence

$$\{X_s, s = 1, 2, \ldots\}$$

then the random sum

$$S = L_1 + L_2 + \cdots + L_N$$

is also independent of the sequence

$$\{e^{iuX_s}, s = 1, 2, \ldots\}.$$

Hence, (3.3.11) can be written in the form

$$\varphi_C(u) = \sum_{s=0}^{\infty} E\big(e^{iuX_1}...e^{iuX_s}\big)P(S = s). \tag{3.3.12}$$

Since the random variables

$$X_1, ..., X_s$$

are independent then the random variables

$$e^{iuX_1}, ..., e^{iuX_s}$$

are independent. Hence (3.3.12) can be written in the form

$$\varphi_C(u) = \sum_{s=0}^{\infty} E\big(e^{iuX_1}\big)...E\big(e^{iuX_s}\big)P(S = s). \tag{3.3.13}$$

We have

$$\varphi_X(u) = E\big(e^{iuX_\kappa}\big), \kappa = 1, 2, ..., s$$

and hence (3.3.13) can be written in the form

$$\varphi_C(u) = \sum_{s=0}^{\infty} \varphi_X^s(u) P(S = s).$$

Since

$$P_S(z) = \sum_{s=0}^{\infty} z^s P(S = s),$$

or equivalently

$$P_S(z) = P_N(P_L(z)),$$

is the probability generating function of the random sum

$$S = L_1 + L_2 + \cdots + L_N$$

then we get that

$$\varphi_C(u) = P_N(P_L(\varphi_X(u))) \tag{3.3.14}$$

is the characteristic function of the random sum

$$C = X_1 + X_2 + \cdots + X_S.$$

The establishment of the stochastic multiplicative model

$$Y = CW$$

as a random contraction of the random sum

$$C = X_1 + X_2 + \cdots + X_S$$

via the random variable

$$W$$

requires the establishment of the independence of the random sum

$$C = X_1 + X_2 + \cdots + X_S$$

and the random variable

$$W.$$

This means the establishment of the relationship

$$\varphi_{C,W}(u,\pi) = \varphi_C(u)\varphi_W(\pi) \tag{3.3.15}$$

where

$$\varphi_{C,W}(u,\pi)$$

is the joint characteristic function of the random sum

$$C = X_1 + X_2 + \cdots + X_S$$

and the random variable

$$W,$$

where

$$\varphi_C(u)$$

is the characteristic function of the random sum

$$C = X_1 + X_2 + \cdots + X_S$$

and

$$\varphi_W(\pi)$$

is the characteristic function of the random variable

$$W.$$

We have

$$\varphi_{C,W}(u,\pi) = E\left(e^{iuC+i\pi W}\right),$$

or equivalently

$$\varphi_{C,W}(u,\pi) = \sum_{s=0}^{\infty} E\left(e^{iuC+i\pi W}|S=s\right)P(S=s). \tag{3.3.16}$$

From (3.3.16) it follows that

$$\varphi_{C,W}(u,\pi)=\sum_{s=0}^{\infty}E\left[e^{iu(X_1+\cdots+X_S)+i\pi W}|S=s\right]P(S=s),$$

or equivalently

$$\varphi_{C,W}(u,\pi)=\sum_{s=0}^{\infty}E\left(e^{iuX_1}...e^{iuX_s}e^{i\pi W}|S=s\right)P(S=s). \tag{3.3.17}$$

Now, we must establish that the random variables

$$X_1, ..., X_s, S, W$$

are independent. This means the establishment of the relationship

$$\varphi_{X_1,...,X_s,S,W}(u_1,u_2,...,u_s,\xi,\pi)=\varphi_{X_1}(u_1)...\varphi_{X_s}(u_s)\varphi_S(\xi)\varphi_W(\pi) \tag{3.3.18}$$

where

$$\varphi_{X_1,...,X_s,S,W}(u_1,u_2,...,u_s,\xi,\pi)$$

is the joint characteristic function of the random variables

$$X_1, ..., X_s, S, W,$$

$$\varphi_{X_\kappa}(u_\kappa)$$

is the characteristic function of the random variable

$$X_\kappa, \kappa=1,2,...,s,$$

$$\varphi_S(\xi)$$

is the characteristic function of the random sum

$$S=L_1+L_2+\cdots+L_N$$

and

$$\varphi_W(\pi)$$

is the characteristic function of the random variable

$$W.$$

We have

$$\varphi_{X_1,\ldots,X_s,S,W}(u_1, u_2, \ldots, u_s, \xi, \pi) = E\big(e^{iu_1X_1+\cdots+iu_sX_s+i\xi S+i\pi W}\big)$$

or equivalently

$$\varphi_{X_1,\ldots,X_s,S,W}(u_1, u_2, \ldots, u_s, \xi, \pi) = \sum_{n=0}^{\infty} E\big(e^{iu_1X_1+\cdots+iu_sX_s+i\xi S+i\pi W}|N = n\big)P(N = n). \tag{3.3.19}$$

From (3.3.19) it follows that

$$\varphi_{X_1,\ldots,X_s,S,W}(u_1, u_2, \ldots, u_s, \xi, \pi) = \sum_{n=0}^{\infty} E\big(e^{iu_1X_1+\cdots+iu_sX_s+i\xi L_1+\cdots+i\xi L_n+i\pi W}|N = n\big)P(N = n). \tag{3.3.20}$$

From the assumption that

$$N,$$

$$\{L_n, n = 1, 2, \ldots\},$$

$$\{X_s, s = 1, 2, \ldots\},$$

and

$$W$$

are independent, it follows the independence of the random variables

$$X_1, \ldots X_s, L_1, \ldots, L_n, W, N.$$

Hence the random variables

$$e^{iu_1X_1}, \ldots, e^{iu_sX_s}, e^{i\xi L_1}, \ldots, e^{i\xi L_n}, e^{i\pi W}, N$$

are also independent. From (3.3.20) it follows that

$$\varphi_{X_1,\ldots,X_s,S,W}(u_1, u_2, \ldots, u_s, \xi, \pi) = E\big(e^{iu_1X_1}\big)\ldots E\big(e^{iu_sX_s}\big)E\big(e^{i\pi W}\big)\sum_{n=0}^{\infty} E\big(e^{i\xi L_1}\big)\ldots E\big(e^{i\xi L_n}\big)P(N = n). \tag{3.3.21}$$

It is readily seen that (3.3.21) can be written in the form

$$\varphi_{X_1,\dots,X_S,S,W}(u_1,u_2,\dots,u_S,\xi,\pi)=\varphi(u_1)\dots\varphi_{X_S}(u_\kappa)\varphi_S(\xi)\varphi_W(\pi). \tag{3.3.22}$$

From (3.3.18) or (3.3.22) it follows that the random variables

$$X_1,\dots,X_S,S,W$$

are independent. Hence (3.3.17) can be written in the form

$$\varphi_{C,W}(u,\pi)=\sum_{s=0}^{\infty}E\big(e^{iu_1X_1}\dots e^{iu_sX_s}e^{i\pi W}\big)P(S=s). \tag{3.3.23}$$

Since the random variables

$$X_1,\dots,X_s,W,$$

are independent then the random variables

$$e^{iu_1X_1},\dots e^{iu_sX_s},e^{i\pi W}$$

are independent. Hence (3.3.23) can be written in the form

$$\varphi_{C,W}(u,\pi)=E\big(e^{i\pi W}\big)\sum_{s=0}^{\infty}E\big(e^{iu_1X_1}\big)\dots E\big(e^{iu_sX_s}\big)P(S=s)$$

or equivalently, (3.3.23) can be written in the form

$$\varphi_{C,W}(u,\pi)=\varphi_W(\pi)\sum_{s=0}^{\infty}\varphi_X^s(u)P(S=s). \tag{3.3.24}$$

Since

$$P_S(\varphi_X(u))=\sum_{s=0}^{\infty}\varphi_X^s(u)P(S=s)$$

then (3.3.24) can be written in the form

$$\varphi_{C,W}(u,\pi)=\varphi_W(\pi)P_S(\varphi_X(u)). \tag{3.3.25}$$

Moreover, since

$$\varphi_C(u)=P_S(\varphi_X(u))$$

then (3.3.25) can be written in the form

$$\varphi_{C,W}(u, \pi) = \varphi_C(u)\varphi_W(\pi). \tag{3.3.26}$$

From (3.3.26) it follows that the positive random sum

$$C = X_1 + X_2 + \cdots + X_S$$

and the random variable

$$W$$

are independent. Moreover, from the fact that the random variable

$$W$$

takes values in the interval

$$(0, 1)$$

it follows that the stochastic multiplicative model

$$Y = CW$$

is a random contraction of the positive random sum

$$C = X_1 + X_2 + \cdots + X_S$$

via the random variable

$$W.$$

3.4 Extending the Contraction of a Random Sum

It is easily seen that an explicit analytical evaluation of the distribution function

$$F_Y(y)$$

of the random contraction of the random sum

$$C = X_1 + X_2 + \cdots + X_S$$

via the random variable

$$W$$

is extremely difficult. For this reason, the present section is devoted to the evaluation of the corresponding characteristic function

$$\varphi_Y(u).$$

Theorem 3.4.1 *Let*

$$F_W(w) \tag{3.4.1}$$

be the distribution function of the random variable

$$W,$$

then the characteristic function of the random contraction

$$Y = CW$$

of the random sum

$$C = X_1 + X_2 + \cdots + X_S,$$

where

$$S = L_1 + L_2 + \cdots + L_N,$$

via the random variable

$$W$$

has the form

$$\varphi_Y(u) = \int_0^1 P_N(P_L(\varphi_X(uw)))dF_W(w) \tag{3.4.2}$$

Proof From (3.3.14) we get that the characteristic function of the positive random sum

$$C = X_1 + X_2 + \cdots + X_S,$$

has the form

$$\varphi_C(u) = P_N(P_L(\varphi_X(u))) \tag{3.4.3}$$

Since the random sum

$$C = X_1 + X_2 + \cdots + X_S$$

is independent of the random variable

$$W,$$

then from (2.8.1), (3.4.1) and (3.4.3) it follows that the characteristic function of the random contraction

$$Y = CW$$

is given by

$$\varphi_Y(u) = \int_0^1 P_N(P_L(\varphi_X(uw)))dF_W(w).$$

The consideration of particular cases of the above characteristic function seems to be of some practical importance.

- We suppose that the random variable

$$N$$

follows the geometric distribution with probability generating function

$$P_N(z) = \frac{pz}{1 - qz}, 0 < p < 1, q = 1 - p,$$

the random variable

$$L$$

follows the geometric distribution with probability generating function

$$P_L(z) = \frac{\omega z}{1 - \zeta z}, 0 < \omega < 1, \zeta = 1 - \omega,$$

the random variable

$$X$$

follows the exponential distribution with characteristic function

$$\varphi_X(u) = \frac{\mu}{\mu - iu}, \mu > 0$$

and the random variable

$$W$$

follows the uniform distribution with distribution function

$$F_W(w) = w, 0 < w < 1.$$

Under these assumptions, the characteristic function

$$\varphi_C(u)$$

in (3.4.2) has the form

$$\varphi_C(u) = \int_0^1 \frac{p\omega\mu}{p\omega\mu - iuw} dw,$$

or equivalently the form

$$\varphi_C(u) = \frac{p\omega\mu}{iu} log\left(\frac{p\omega\mu}{p\omega\mu - iu}\right).$$

- We suppose that the random variable

$$N$$

follows the Pascal distribution with probability generating function

$$P_N(z) = \left(\frac{pz}{1 - qz}\right)^2, 0 < p < 1, q = 1 - p,$$

the random variable

$$L$$

follows the geometric distribution with probability generating function

$$P_L(z) = \frac{\omega z}{1 - \zeta z}, 0 < \omega < 1, \zeta = 1 - \omega,$$

the random variable

$$X$$

follows the exponential distribution with characteristic function

$$\varphi_X(u) = \frac{\mu}{\mu - iu}, \mu > 0,$$

and the random variable

$$W$$

follows the uniform distribution with distribution function

$$F_W(w) = 1, 0 < w < 1.$$

Under these assumptions, the characteristic function

$$\varphi_C(u)$$

in (3.4.2) has the form

$$\varphi_C(u) = \int_0^1 \left(\frac{p\omega\mu}{p\omega\mu - iuw} \right)^2 dw,$$

or equivalently the form

$$\varphi_C(u) = \frac{p\omega\mu}{p\omega\mu - iu}.$$

- We suppose that the random variable

$$N$$

follows the Pascal distribution with probability generating function

$$P_N(z) = \left(\frac{pz}{1 - qz}\right)^{\kappa+1},$$

where

$$0 < p < 1,$$

$$q = 1 - p,$$

and

$$\kappa$$

positive integer, the random variable

$$L$$

follows the geometric distribution with probability generating function

$$P_L(z) = \frac{\omega z}{1 - \zeta z}, 0 < \omega < 1, \zeta = 1 - \omega,$$

the random variable

$$X$$

follows the exponential distribution with characteristic function

$$\varphi_X(u) = \frac{\mu}{\mu - iu}, \mu > 0,$$

and the random variable

$$W$$

follows the beta distribution with distribution function

$$F_W(w) = w^k, 0 < w < 1.$$

Under these assumptions, the characteristic function

$$\varphi_C(u)$$

in (3.4.2) has the form

$$\varphi_C(u) = \int_0^1 \left(\frac{p\omega\mu}{p\omega\mu - iuw} \right)^{\kappa+1} w^{\kappa-1} dw. \tag{3.4.4}$$

From (2.15.9) it follows that the characteristic function in (3.4.4) can be written in the form
$\varphi_C(u) = \left(\frac{p\omega\mu}{p\omega\mu - iu} \right)^{\kappa}$.

- We suppose that the random variable

$$N$$

follows the geometric distribution with probability generating function

$$P_N(z) = \frac{pz}{1 - qz}, 0 < p < 1, q = 1 - p$$

the random variable

$$L$$

follows the geometric distribution with probability generating function

$$P_L(z) = \frac{\omega z}{1 - \zeta z}, 0 < \omega < 1, \zeta = 1 - \omega,$$

the random variable

$$X$$

follows the exponential distribution with characteristic function

$$\varphi_X(u) = \frac{\mu}{\mu - iu}, \mu > 0,$$

and the random variable

$$W$$

follows the beta distribution with distribution function

$$F_W(w) = w^2, 0 < w < 1.$$

Under these assumptions, the characteristic function

$$\varphi_C(u)$$

in (3.4.2) has the form

$$\varphi_C(u) = 2\int_0^1 \frac{p\omega\mu}{p\omega\mu - iuw} w dw. \tag{3.4.5}$$

Integrating in (3.4.5) we get that

$$\varphi_C(u) = \frac{\frac{p\omega\mu}{iu}\log\left(\frac{p\omega\mu}{p\omega\mu - iu}\right) - 1}{{}^{iu}\!/_{2p\omega\mu}}. \tag{3.4.6}$$

- We suppose that the random variable

$$N$$

follows the Pascal distribution with probability generating function

$$P_N(z) = \left(\frac{pz}{1-qz}\right)^2, 0 < p < 1, q = 1-p,$$

the random variable

$$L$$

follows the geometric distribution with probability generating function

$$P_L(z) = \frac{\omega z}{1-\zeta z}, 0 < \omega < 1, \zeta = 1-\omega,$$

the random variable

$$X$$

follows the exponential distribution with characteristic function

$$\varphi_X(u) = \frac{\mu}{\mu - iu}, \mu > 0,$$

and the random variable

$$W$$

follows the beta distribution with distribution function

$$F_W(w) = 1 - (1 - w)^2, 0 < w < 1.$$

Under these assumptions, the characteristic function

$$\varphi_C(u)$$

in (3.4.2) has the form

$$\varphi_C(u) = 2 \int_0^1 \left(\frac{p\omega\mu}{p\omega\mu - iuw} \right)^2 (1 - w)dw. \tag{3.4.7}$$

Integrating in (3.4.7) we get that

$$\varphi_C(u) = \frac{\frac{p\omega\mu}{iu} \log\left(\frac{p\omega\mu}{p\omega\mu - iu}\right) - 1}{{}^{iu}/_{2p\omega\mu}}.$$

The above characteristic function coincides with the characteristic function in (3.4.6).

3.5 Interpretation of the Formulated Random Contraction in Stochastic Discounting

The present section is devoted to the establishment of the formulated random contraction of Sect. 3.3 as stochastic discounting model.

It is shown that such a contraction can be represented as the present value of a random cash flow arising at a future random time. More precisely, the section provides sufficient conditions in order the stochastic discounting model

$$Y = Ce^{-\delta T},$$

where

$$C = X_1 + X_2 + \cdots + X_S,$$

$$S = L_1 + L_2 + \cdots + L_N,$$

$$\delta$$

a positive real number, and

$$T$$

a positive random variable, to be a random contraction. The above stochastic discounting model denotes the present value of a random cash flow under random time and constant force of interest.

Theorem 3.5.1 *Let*

$$N$$

be a discrete random variable with values in the set

$$\mathbf{N}_0$$

and let

$$\{L_n n = 1, 2, ...\}$$

be a sequence of discrete independent random variables with values in the set

$$\mathbf{N}_0.$$

We suppose that the random variables of the sequence

$$\{L_n n = 1, 2, ...\}$$

are equally distributed with the random variable

$$L,$$

and we consider the random variable

$$S = L_1 + L_2 + \cdots + L_N.$$

Let

$$\{\mathrm{X}_s s = 1, 2, ...\}$$

be a sequence of positive independent random variables. We suppose that the random variables of the sequence

$$\{\mathrm{X}_s s = 1, 2, ...\}$$

are equally distributed with the random variable

$$X.$$

We set

$$C = X_1 + X_2 + \cdots + X_S$$

and let

$$T,$$

be a positive random variable. Moreover, we suppose that

$$\delta$$

is a positive real number and we consider the random variable

$$W = e^{-\delta T}$$

with values in the interval

$$(0, 1).$$

If

$$N,$$

$$\{L_n n = 1, 2, ...\},$$

$$\{\mathrm{X}_s s = 1, 2, ...\},$$

and

$$T$$

are independent then the stochastic multiplicative model

$$Y = Ce^{-\delta T},$$

or equivalently the stochastic multiplicative model

$$Y = CW,$$

is a random contraction of the random variable

$$C = X_1 + X_2 + \cdots + X_S$$

via the random variable

$$W = e^{-\delta T}.$$

Proof From Theorem 3.3.1 it follows that the random sum

$$C = X_1 + X_2 + \cdots + X_S$$

is independent of the random variable

$$T.$$

Hence the random sum

$$C = X_1 + X_2 + \cdots + X_S$$

is also independent of the random

$$W = e^{-\delta T}.$$

From the fact that the random variable

$$W = e^{-\delta T}$$

takes values in the interval

$$(0, 1)$$

we conclude that the stochastic multiplicative model

$$Y = Ce^{-\delta T},$$

or equivalently the stochastic multiplicative model

$$Y = CW$$

is a contraction of the random sum

$$C = X_1 + X_2 + \cdots + X_S$$

via the random variable

$$W = e^{-\delta T}.$$

An interpretation of the stochastic multiplicative model

$$Y = Ce^{-\delta T}$$

in the area of continuous discounting is the following.

We suppose that the random variable

$$N$$

denotes the number of different groups of cash flows arising at the random time point

$$T.$$

We also suppose that the random variable

$$L_n, n = 1, 2, ...$$

denotes the number of cash flows consisting the

$$n\text{th}$$

group of cash flows. Hence the random variable

$$S = L_1 + L_2 + \cdots + L_N$$

denotes the total number of cash flows which are incorporated by the

$$N$$

different groups of cash flows arising at the random time point

$$T.$$

Moreover, we suppose that the random variable

$$\mathrm{X}_s s = 1, 2, \ldots$$

denotes the size of the

$$s\text{th}$$

cash flow at the random time point

$$T.$$

Hence the random variable

$$C = X_1 + X_2 + \cdots + X_S$$

denotes the size of the

$$S = L_1 + L_2 + \cdots + L_N$$

cash flows arising at the random time point

$$T.$$

If the positive real number

$$\delta$$

denotes force of interest then the stochastic multiplicative model

$$Y = Ce^{-\delta T}$$

denotes the present value at the time point

$$0$$

of the cash flow

$$C = X_1 + X_2 + \cdots + X_S$$

arising at the random time

$$T.$$

The theoretical contribution of the present section consists of the interpretation of the formulated random contraction as a stochastic discounting model. The practical contribution of that section consists of making quite clear the vital applicability of such a stochastic model in implementing proactive operations. A significant part of the third chapter of the present work concentrates on the role of these contributions in supporting proactivity in global information risk management operations.

3.6 Characteristic Function of the Formulated Stochastic Discounting Model

It is quite obvious that an explicit analytical evaluation of the distribution function

$$F_Y(y)$$

of the random contraction of the random sum

$$C = X_1 + X_2 + \cdots + X_S$$

via the random variable

$$W = e^{-\delta T},$$

or equivalently of the stochastic discounting model

$$Y = Ce^{-\delta T},$$

is extremely difficult. For this reason, the present section is devoted to the evaluation of the corresponding characteristic function

$$\varphi_Y(u).$$

Theorem 3.6.1 *Let*

$$F_T(t) \tag{3.6.1}$$

be the distribution function of the random variable

$$T$$

and

$$F_W(w)$$

be the distribution function of the random variable

$$W = e^{-\delta T},$$

the characteristic function of the random contraction

$$Y = CW$$

of the positive random sum

$$C = X_1 + X_2 + \cdots + X_S$$

where

$$S = L_1 + L_2 + \cdots + L_N,$$

via the random variable

$$W = e^{-\delta T},$$

or equivalently the characteristic function of the stochastic discounting model

$$Y = Ce^{-\delta T}$$

has the form

$$\varphi_Y(u) = \int_0^1 P_N(P_L(\varphi_X(uw)))dF_W(w),$$

or equivalently

$$\varphi_Y(u) = \int_0^1 P_N(P_L(\varphi_X(uw)))d\left[1 - F_T\left(-\frac{1}{\delta}\log w\right)\right].$$

Proof From (3.3.14) we get that the characteristic function of the random sum

$$C = X_1 + X_2 + \cdots + X_S$$

has the form

$$\varphi_C(u) = P_N(P_L(\varphi_X(u))). \tag{3.6.2}$$

From (3.6.1) it readily follows that the distribution function of the random variable

$$W = e^{-\delta T}$$

is given by

$$F_W(w) = 1 - F_T\left(-\frac{1}{\delta}\log w\right) \tag{3.6.3}$$

Since the random sum

$$C = X_1 + X_2 + \cdots + X_S$$

is independent of the random variable

$$W = e^{-\delta T}$$

then from (2.8.1), (3.6.1), (3.6.2) and (3.6.3) it follows that the characteristic function of the random contraction

$$Y = CW$$

is given by

$$\varphi(u) = \int_0^1 P_N(P_L(\varphi_X(uw)))dF_W(w),$$

or equivalently the characteristic function of the stochastic discounting model

$$Y = Ce^{-\delta T}$$

is given by

$$\varphi_Y(u) = \int_0^1 P_N(P_L(\varphi_X(uw)))d\left[1 - F_T\left(-\frac{1}{\delta}\log w\right)\right]. \tag{3.6.4}$$

Below we consider a particular case of the above characteristic function which is of some practical and theoretical importance.

- We suppose that the random variable

$$N$$

is arbitrarily distributed with probability generating function

$$P_N(z),$$

the random variable

$$L$$

is arbitrarily distributed with probability generating function

$$P_L(z),$$

the random variable

$$X$$

is arbitrarily distributed with characteristic function

$$\varphi_X(u)$$

and the random variable

$$T$$

is exponentially distributed with distribution function

$$F_T(t) = 1 - e^{-\mu t}, \mu > 0, t \geq 0. \tag{3.6.5}$$

From (3.6.3) and (3.6.5) it follows that the distribution function of the random variable

$$W = e^{-\delta T}$$

is given by

$$F_W(w) = w^a, a > 0, 0 < w < 1, \tag{3.6.6}$$

where

$$a = {}^{\lambda}/_{\delta}.$$

Moreover (3.6.2) and (3.6.6) imply that the characteristic function

$$\varphi_Y(u)$$

in (3.6.4) corresponding to the random contraction

$$Y = CW$$

or equivalently corresponding to the stochastic discounting model

$$Y = Ce^{-\delta T}$$

is given by

$$\varphi_Y(u) = a \int_0^1 P_N(P_L(\varphi_X(uw)))w^{a-1}dw. \tag{3.6.7}$$

From Sect. 2.10 it follows that the characteristic function

$$\varphi_Y(u)$$

in (3.6.7) corresponds to a member of the class of a-unimodal random contractions.

A significant part of the third chapter of the present work makes use of the theoretical and practical results of this section, the general recognition of the exponential distribution as the most important distribution of time, and the consideration of proactivity as a structural element of intelligent behaviour of systems in order to investigate the role of the proposed interpretation of the formulated random contraction, as a stochastic discounting model, in global risk governance operations.

3.7 Extension of a Stochastic Multiplicative Model

The present section introduces an extension of the stochastic multiplicative model of Sect. 3.2 by making use of a discrete random variable, a sequence of discrete independent and identically distributed random variables, a sequence of positive independent and identically distributed random variables, a Bernoulli random variable, and two positive random variables.

We consider the discrete random variable

$$N$$

with values in the set

$$\boldsymbol{N}_0,$$

the sequence

$$\{L_n n = 1, 2, ...\}$$

of discrete independent and identically distributed random variables with values in the set

$$\boldsymbol{N}_0$$

and the discrete random variable

$$S = L_1 + L_2 + \cdots + L_N.$$

We also consider the sequence

$$\{X_s, s = 1, 2, ...\},$$

of positive independent and identically distributed random variables, the positive random variable

$$C = X_1 + X_2 + \cdots + X_S,$$

the positive random variable

$$B,$$

the random variable

$$W$$

with values in the interval

$$(0, 1),$$

and the Bernoulli random variable

$$I.$$

Moreover, we consider the random variable

$$J = \begin{cases} B \ , \ I = 0 \\ C \ , \ I = 1 \end{cases}$$

and the stochastic multiplicative model

$$U = JW.$$

The practical and theoretical applications of the above stochastic multiplicative model are mainly based on the discrete random variable

$$S = L_1 + L_2 + \cdots + L_N,$$

the positive random variable

$$C = X_1 + X_2 + \cdots + X_S, ,$$

and the positive random variable

$$J = \begin{cases} B\ , \ I = 0 \\ C\ , \ I = 1 \end{cases}$$

The incorporation of a Bernoulli random variable in extending the stochastic multiplicative model of Sect. 3.2 is readily recognized as a very significant factor for the description, analysis, and solution of problems arising in a wide variety of practical disciplines. Moreover, the incorporated Bernoulli random variable substantially facilitates the investigation of the probability distribution corresponding to the extension of the stochastic multiplicative model.

In addition the representation as a random contraction of such an extension is of particular theoretical and practical importance in many significant theoretical and practical disciplines.

3.8 Representation as a Random Contraction of an Extension of a Stochastic Multiplicative Model

The establishment of sufficient conditions for representing the stochastic multiplicative model

$$U = JW$$

as a random contraction of the random variable

$$J = \begin{cases} B\ , \ I = 0 \\ C\ , \ I = 1 \end{cases}$$

via the random variable

$$W$$

is the purpose of the present section.

Theorem 3.8.1 *Let*

$$N$$

be a discrete random variable with values in the set

$$\boldsymbol{N}_0$$

and probability generating function

$$P_N(z)$$

and let

$$\{L_n, n = 1, 2, ...\}$$

be a sequence of discrete independent random variables with values in the set

$$\boldsymbol{N}_0.$$

We suppose that the random variables of the sequence

$$\{L_n, n = 1, 2, ...\}$$

are equally distributed with the random variable

$$L$$

having probability generating function

$$P_L(z),$$

and we consider the random variable

$$S = L_1 + L_2 + \cdots + L_N.$$

Let

$$\{X_s, s = 1, 2, ...\}$$

be a sequence of positive independent random variables. We suppose that the random variables of the sequence

$$\{X_s, s = 1, 2, ...\}$$

are equally distributed with the random variable

$$X$$

having characteristic function

$$\varphi_X(u).$$

Let

$$B$$

be a positive random variable with characteristic function

$$\varphi_B(u),$$

$$I,$$

a Bernoulli random variable with probability generating function

$$P_I(z) = 1 - d = dz, 0 < d < 1$$

and

$$W$$

a random variable with characteristic function

$$\varphi_W(\pi)$$

and values in the interval

$$(0, 1).$$

If

$$\mathrm{N},$$

$$\{L_n, n = 1, 2, ...\},$$

$$\{X_s, s = 1, 2, ...\},$$

$$B,$$

$$I,$$

and

$$W$$

are independent then the stochastic multiplicative model

$$U = JW$$

is a random contraction of the random variable

$$J = \begin{cases} B\ , I = 0 \\ C\ , I = 1 \end{cases}$$

via the random variable

$$W.$$

Proof The establishment of the random variable

$$S = L_1 + L_2 + \cdots + L_N$$

as a random sum with probability generating function

$$P_S(z) = P_N(P_L(z))$$

and the establishment of the random variable

$$C = X_1 + X_2 + \cdots + X_S$$

as a random sum with characteristic function

$$\varphi_C(u) = P_N(P_L(\varphi_X(u)))$$

are given by *Theorem 3.3.1*. Moreover, the establishment of the stochastic multiplicative model

$$U = JW$$

as a random contraction of the random variable

$$J = \begin{cases} B\ , I = 0 \\ C\ , I = 1 \end{cases}$$

via the random variable

$$W$$

requires the establishment of the independence of the random variable

$$J = \begin{cases} B\ , \ I = 0 \\ C\ , \ I = 1 \end{cases}$$

and the random variable

$$W.$$

This means the establishment of the relationship

$$\varphi_{J,W}(u, \pi) = \varphi_J(u)\varphi_W(\pi), \tag{3.8.1}$$

where

$$\varphi_{J,W}(u, \pi)$$

is the joint characteristic function of the random variables

$$J = \begin{cases} B\ , \ I = 0 \\ C\ , \ I = 1 \end{cases}$$

and

$$W,$$

$$\varphi_J(u)$$

is the characteristic function of the random variable

$$J = \begin{cases} B\ , \ I = 0 \\ C\ , \ I = 1 \end{cases}$$

and

$$\varphi_W(\pi)$$

is the characteristic function of the random variable

$$W.$$

We have

$$\varphi_{J,W}(u,\pi) = E\big(e^{iuJ+i\pi W}\big)$$

or equivalently

$$\varphi_{J,W}(u,\pi) = E\big(e^{iuJ+i\pi W}|I=0\big)P(I=0) + E\big(e^{iuJ+i\pi W}|I=1\big)P(I=1). \tag{3.8.2}$$

From (3.8.2) it follows that

$$\varphi_{J,W}(u,\pi) = (1-d)E\big(e^{iuB+i\pi W}|I=0\big) + dE\big(e^{iuC+i\pi W}|I=1\big). \tag{3.8.3}$$

Since the random variables

$$B, W, I,$$

are independent then (3.8.3) can be written in the form

$$\varphi_{J,W}(u,\pi) = (1-d)\varphi_{\mathrm{B}}(u)\varphi_{W}(\pi) + dE\big(e^{iuC+i\pi W}|I=1\big). \tag{3.8.4}$$

We shall prove the independence of the random variables

$$C, W, I.$$

This requires the establishment of the relationship

$$\varphi_{C,W,I}(u,\pi,\eta) = \varphi_C(u)\varphi_W(\pi)\varphi_I(\eta)$$

where

$$\varphi_{C,W,I}(u,\pi,\eta)$$

is the joint characteristic function of the random variables

$$C = X_1 + X_2 + \cdots + X_S,$$

$$W$$

and

$$I,$$

$$\varphi_C(u) = P_N(P_L(\varphi_X(u))),$$

is the characteristic function of the random variable

$$C = X_1 + X_2 + \cdots + X_S,$$

$$\varphi_W(\pi)$$

is the characteristic function of the random variable

$$W$$

and

$$\varphi_I(\eta)$$

is the characteristic function of the random variable

$$I.$$

We have

$$\varphi_{C,W,I}(u, \pi, \eta) = E\big(e^{iuC+i\pi W+i\eta I}\big)$$

or equivalently

$$\varphi_{C,W,I}(u, \pi, \eta) = \sum_{s=0}^{\infty} E\big(e^{iuC+i\pi W+i\eta I}|S = s\big)P(S = s). \tag{3.8.5}$$

From (3.8.5) it follows that

$$\varphi_{C,W,I}(u, \pi, \eta) = \sum_{s=0}^{\infty} E\big(e^{iu(X_1+X_2+\cdots+X_S)+i\pi W+i\eta I}|S = s\big)P(S = s)$$

or equivalently

$$\varphi_{C,W,I}(u, \pi, \eta) = \sum_{s=0}^{\infty} E\big(e^{iuX_1}...e^{iuX_s}e^{i\pi W}e^{i\eta I}|S = s\big)P(S = s). \tag{3.8.6}$$

Now we must establish that the random variables

$$X_1, X_2, ..., X_s, S, W, I$$

are independent. This means the establishment of the relationship

$$\varphi_{X_1,\ldots,X_s,S,W,I}(u_1, \ldots, u_s, \xi, \pi, \eta) = \varphi_{X_1}(u_1)\ldots\varphi_{X_s}(u_s)\varphi_S(\xi)\varphi_W(\pi)\varphi_I(\eta) \quad (3.8.7)$$

where

$$\varphi_{X_1,\ldots,X_s,S,W,I}(u_1, \ldots, u_s, \xi, \pi, \eta)$$

is the joint characteristic function of the random variables

$$X_1, X_2, \ldots, X_s, S, W, I,$$

$$\varphi_{X_\kappa}(u_\kappa).$$

is the characteristic function of the random variable

$$X_\kappa, \kappa = 1, 2, \ldots, s,$$

$$\varphi_S(\xi),$$

is the characteristic function of the random sum

$$S = L_1 + L_2 + \cdots + L_N$$

$$\varphi_W(\pi),$$

is the characteristic function of the random variable

$$W$$

and

$$\varphi_I(\eta)$$

is the characteristic function of the random variable

$$I.$$

We have

$$\varphi_{X_1,\ldots,X_s,S,W,I}(u_1, \ldots, u_s, \xi, \pi, \eta) = E\left(e^{iu_1X_1+\cdots+iu_sX_s+i\xi S+i\pi W+i\eta I}\right)$$

or equivalently

,

$$\varphi_{X_1,\ldots,X_s,S,W,I}(u_1,\ldots,u_s,\xi,\pi,\eta)$$
$$=\sum_{n=0}^{\infty}E\left(e^{iu_1X_1+\cdots+iu_sX_s+i\xi S+i\pi W+i\eta I}|N=n\right)P(N=n). \tag{3.8.8}$$

From (3.8.8) it follows that

$$\varphi_{X_1,\ldots,X_s,S,W,I}(u_1,\ldots,u_s,\xi,\pi,\eta)$$
$$=\sum_{n=0}^{\infty}E\left(e^{iu_1X_1+\cdots+iu_sX_s+i\xi L_1+\cdots+i\xi L_n+i\pi W+i\eta I}|N=n\right)P(N=n). \tag{3.8.9}$$

The independence of

$$N.$$

$$\{L_n n=1,2,\ldots\}$$

$$\{X_s,s=1,2,\ldots\}$$

$$W,$$

and

$$I$$

implies the independence of the random variables

$$X_1,\ldots,X_s,L_1,\ldots,L_n,W,I,N.$$

Hence the random variables

$$e^{iu_1X_1},\ldots,e^{iu_sX_s},e^{i\xi L_1},\ldots,e^{i\xi L_n},e^{i\pi W},e^{i\eta I},N$$

are also independent. From (3.8.9) it follows that

$$\varphi_{X_1,\ldots,X_s,S,W,I}(u_1,\ldots,u_s,\xi,\pi,\eta)$$
$$=E\left(e^{iuX_1}\right)\ldots E\left(e^{iuX_s}\right)E\left(e^{i\pi W}\right)E\left(e^{i\eta I}\right)\sum_{n=0}^{\propto}E\left(e^{i\xi L_1}\right)\ldots E\left(e^{i\xi L_n}\right)P(N=n)$$

or equivalently

$$\varphi_{X_1,...,X_s,S,W,I}(u_1, ..., u_s, \xi, \pi, \eta) = \varphi_{X_1}(u_1)...\varphi_{X_s}(u_s)P_N(\phi_L(\xi))\varphi_W(\pi)\varphi_I(\eta). \tag{3.8.10}$$

Since

$$\varphi_S(\xi) = P_N(\varphi_L(\xi))$$

is the characteristic function of the random sum

$$S = L_1 + L_2 + \cdots + L_N$$

then (3.8.10) can be written in the form

$$\varphi_{X_1,...,X_s,S,W,I}(u_1, ..., u_s, \xi, \pi, \eta) = \varphi_{X_1}(u_1)...\varphi_{X_s}(u_s)\varphi_S(\xi)\varphi_W(\pi)\varphi_I(\eta) \tag{3.8.11}$$

From (3.8.11) it follows that the random variables

$$X_1, X_2, ..., X_s, S, W, I$$

are independent. Hence (3.8.6) can be written in the form

$$\varphi_{C,W,I}(u, \pi, \eta) = \sum_{s=0}^{\infty} E\left(e^{iu_1X_1}...e^{iu_sX_s}e^{i\pi W}e^{i\eta I}\right)P(S = s). \tag{3.8.12}$$

Since the random variables

$$X_1, X_2, ..., X_s, W, I$$

are independent, the random variables

$$e^{iu_1X_1}, ..., e^{iu_sX_s}, e^{i\pi W}, e^{i\eta I}$$

are also independent. Hence (3.8.12) can be written in the form

$$\varphi_{C,W,I}(u, \pi, \eta) = E\left(e^{i\pi W}\right)E\left(e^{i\eta I}\right)\sum_{s=0}^{\infty} E\left(e^{iuX_1}\right)...E\left(e^{iu_sX_s}\right)P(S = s)$$

or equivalently in the form

$$\varphi_{C,W,I}(u, \pi, \eta) = \varphi_W(\pi)\varphi_I(\eta)E\left(e^{i\eta I}\right)\sum_{s=0}^{\infty} \varphi_X^s(u)P(S = s) \tag{3.8.13}$$

Since

$$P_S(\varphi_X(u)) = \sum_{s=0}^{\infty} \varphi_X^s(u) P(S = s)$$

then (3.8.13) can be written in the form

$$\varphi_{C,W,I}(u, \pi, \eta) = \varphi_W(\pi)\varphi_I(\eta) P_S(\varphi_X(u)). \tag{3.8.14}$$

Moreover, since

$$\varphi_C(u) = P_S(\varphi_X(u))$$

then (3.8.14) can be written in the form

$$\varphi_{C,W,I}(u, \pi, \eta) = \varphi_C(u)\varphi_W(\pi)\varphi_{\mathrm{I}}(\eta)$$

which means that the random variables

$$C, W, I,$$

are independent. From (3.8.4) and the independence of the random variables

$$C, W, I,$$

it follows that

$$\varphi_{J.W}(u, \pi) = (1 - d)\varphi_B(u)\varphi_W(\pi) + d\varphi_C(u)\varphi_W(\pi). \tag{3.8.15}$$

and for the characteristic function of the random variable

$$J = \begin{bmatrix} B \ , \ I = 0 \\ \\ C \ , \ I = 1 \end{bmatrix}$$

we can write

$$\varphi_J(u) = E\big(e^{iuJ}\big)$$

or

$$\varphi_J(u) = E\big(e^{iuB} | I = 0\big) P(I = 0) + E\big(e^{iuC} | I = 1\big) P(I = 1). \tag{3.8.16}$$

From (3.8.16) we get that

$$\varphi_J(u) = (1 - d)\varphi_B(u) + d\varphi_C(u). \tag{3.8.17}$$

Moreover, (3.8.15) and (3.8.17) imply that

$$\varphi_{J,W}(u,\pi) = \varphi_J(u)\varphi_W(\pi). \tag{3.8.18}$$

From (3.8.1) or (3.8.18) it follows that the random variable

$$J = \begin{cases} B\ , I = 0 \\ C\ , I = 1 \end{cases}$$

is independent of the random variable

$$W.$$

Hence the stochastic multiplicative model

$$U = JW,$$

is a random contraction of the random variable

$$J = \begin{cases} B\ , I = 0 \\ C\ , I = 1 \end{cases}$$

via the random variable

$$W.$$

3.9 Characteristic Function of an Extension of a Random Contraction

It is obvious that the random contraction

$$U = JW$$

of Sect. 3.8 is an extension of the random contraction

$$Y = CW,$$

of Sect. 3.3. Since an explicit analytical evaluation of the distribution function

$$F_Y(y)$$

is extremely difficult then Sect. 3.4 evaluates the corresponding characteristic function

$$\varphi_Y(u).$$

Similarly an explicit analytical evaluation of the distribution function

$$F_U(\upsilon)$$

of the random contraction of the random variable

$$J = \begin{bmatrix} B\ , I = 0 \\ C\ , I = 1 \end{bmatrix}$$

via the random variable

$$W$$

is extremely difficult. For this reason the present section is devoted to the evaluation of the corresponding characteristic function

$$\varphi_U(u).$$

Theorem 3.9.1 *Let*

$$F_W(w) \tag{3.9.1}$$

be the distribution function of the random variable

$$W,$$

then the characteristic function of the random contraction

$$U = JW$$

of the random variable

$$J = \begin{bmatrix} B\ , I = 0 \\ C\ , I = 1 \end{bmatrix}$$

via the random variable

$$W$$

has the form

$$\varphi_U(u) = \int_0^1 [(1-\rho)\varphi_B(uw) + \rho\varphi_C(uw)]dF_W(w)$$

or equivalently the form

$$\varphi_U(u) = \int_0^1 [(1-\rho)\varphi_B(uw) + \rho P_N(P_L(\varphi_X(uw)))]dF_W(w). \tag{3.9.2}$$

Proof From (3.3.14) we get that the characteristic function of the random sum

$$C = X_1 + X_2 + \cdots + X_S$$

has the form

$$\varphi_C(u) = P_N(P_L(\varphi_X(u))) \tag{3.9.3}$$

and the from (3.8.17) we get that the characteristic function of the random variable

$$J = \begin{bmatrix} B\ , \ I = 0 \\ \\ C\ , \ I = 1 \end{bmatrix}$$

has the form

$$\varphi_J(u) = (1-\rho)\varphi_B(u) + \rho\varphi_C(u)$$

or equivalently the form

$$\varphi_J(u) = (1-\rho)\varphi_B(u) + \rho P_N(P_L(\varphi_X(u))). \tag{3.9.4}$$

Since the random variable

$$J = \begin{bmatrix} B\ , \ I = 0 \\ \\ C\ , \ I = 1 \end{bmatrix}$$

is independent of the random variable

$$W$$

then (2.8.1), (3.9.1) and (3.9.4) imply that the characteristic function of the random contraction

$$U = JW$$

is given by

$$\varphi_U(u) = \int_0^1 [(1-\rho)\varphi_B(uw) + \rho P_N(P_L(\varphi_X(uw)))] dF_W(w).$$

The consideration of particular cases of the above characteristic function seems to be of some particular importance.

- We suppose that the random variable

$$N$$

follows the geometric distribution with probability generating function

$$P_N(z) = \frac{pz}{1-qz}, 0 < p < 1, q = 1-p,$$

the random variable

$$L$$

follows the geometric distribution with probability generating function

$$P_L(z) = \frac{\omega z}{1-\zeta z}, 0 < \omega < 1, \zeta = 1-\omega,$$

the random variable

$$X$$

follows the exponential distribution with characteristic function

$$\varphi_X(u) = \frac{\mu}{\mu - iu}, \mu > 0,$$

the random variable

$$W$$

follows the uniform distribution with distribution function

$$F_W(w) = w, 0 < w < 1$$

and the random variable

$$B$$

follows the exponential distribution with characteristic function

$$\varphi_B(u) = \frac{\mu}{\mu - iu}, \mu > 0.$$

Under these assumptions the characteristic function

$$\varphi_U(u)$$

in (3.9.2) has the form

$$\varphi_U(u) = (1-\rho)\frac{\mu}{iu}\log\left(\frac{\mu}{\mu - iu}\right) + \rho\frac{p\omega\mu}{iu}\log\left(\frac{p\omega\mu}{p\omega\mu - iu}\right).$$

- We suppose that the random variable

$$N$$

follows the Pascal distribution with probability generating function

$$P_N(z) = \left(\frac{pz}{1-qz}\right)^2, 0 < p < 1, q = 1 - p,$$

the random variable

$$L$$

follows the geometric distribution with probability generating function

$$P_L(z) = \frac{\omega z}{1 - \zeta z}, 0 < \omega < 1, \zeta = 1 - \omega,$$

the random variable

$$X$$

follows the exponential distribution with characteristic function

$$\varphi_X(u) = \frac{\mu}{\mu - iu}, \mu > 0,$$

the random variable

$$W$$

follows the uniform distribution with distribution function

$$F_W(w) = w, 0 < w < 1$$

and the random variable

$$B$$

follows the exponential distribution with characteristic function

$$\varphi_B(u) = \frac{\mu}{\mu - iu}, \mu > 0.$$

Under these assumptions the characteristic function

$$\varphi_U(u)$$

in (3.9.2) has the form

$$\varphi_U(u) = (1-\rho)\frac{\mu}{iu}\log\left(\frac{\mu}{\mu - iu}\right) + \rho\left(\frac{p\omega\mu}{p\omega\mu - iu}\right).$$

- We suppose that the random variable

$$N$$

follows the Pascal distribution with probability generating function

$$P_N(z) = \left(\frac{pz}{1-qz}\right)^{\kappa+1}$$

where

$$0 < p < 1$$

$$q = 1 - p$$

and

$$\kappa$$

positive integer, the random variable

$$L$$

follows the geometric distribution with probability generating function

$$P_L(z) = \frac{\omega z}{1 - \zeta z}, 0 < \omega < 1, \zeta = 1 - \omega,$$

the random variable

$$X$$

follows the exponential distribution with characteristic function

$$\varphi_X(u) = \frac{\mu}{\mu - iu}, \mu > 0,$$

the random variable

$$B$$

follows the gamma distribution with characteristic function

$$\phi_B(u) = \left(\frac{\mu}{\mu - iu}\right)^{\kappa+1},$$

and the random variable

$$W$$

follows the beta distribution with distribution function

$$F_W(w) = w^\kappa, 0 < w < 1.$$

Under these assumptions the characteristic function

$$\varphi_U(u),$$

in (3.9.2) has the form

$$\varphi_U(u) = (1-\rho)\left(\frac{\mu}{\mu - iu}\right)^{\kappa} + \rho\left(\frac{p\omega\mu}{p\omega\mu - iu}\right)^{\kappa}.$$

- We suppose that the random variable

$$N$$

follows the geometric distribution with probability generating function

$$P_N(z) = \frac{pz}{1-qz}, 0 < p < 1, q = 1-p$$

the random variable

$$L$$

follows the geometric distribution with probability generating function

$$P_L(z) = \frac{\omega z}{1-\zeta z}, 0 < \omega < 1, \zeta = 1-\omega$$

the random variable

$$X$$

follows the exponential distribution with characteristic function

$$\varphi_X(u) = \frac{\mu}{\mu - iu}, \mu > 0,$$

the random variable

$$B$$

follows the exponential distribution with characteristic function

$$\varphi_{\mathrm{B}}(u) = \frac{\mu}{\mu - iu}, \mu > 0,$$

and the random variable

$$W$$

follows the beta distribution with distribution function

$$F_W(w) = w^2, 0 < w < 1$$

Under these assumptions the characteristic function

$$\varphi_U(u)$$

in (3.9.2) has the form

$$\varphi_U(u) = (1-\rho)\frac{\frac{\mu}{iu}\log\left(\frac{\mu}{\mu-iu}\right)-1}{{}^{iu}/_{2\mu}} + \rho\frac{\frac{p\omega\mu}{iu}\log\left(\frac{p\omega\mu}{p\omega\mu-iu}\right)-1}{{}^{iu}/_{2p\mu}}.$$

- We suppose that the random variable

$$N$$

follows the Pascal distribution with probability generating function

$$P_N(z) = \left(\frac{pz}{1-qz}\right)^2, 0 < p < 1, q = 1-p$$

the random variable

$$L$$

follows the geometric distribution with probability generating function

$$P_L(z) = \frac{\omega z}{1-\zeta z}, 0 < \omega < 1, \zeta = 1-\omega$$

the random variable

$$X$$

follows the exponential distribution with characteristic function

$$\varphi_X(u) = \frac{\mu}{\mu - iu}, \mu > 0,$$

the random variable

$$B$$

follows the exponential distribution with characteristic function

$$\varphi_B(u) = \left(\frac{\mu}{\mu - iu}\right)^2, \mu > 0$$

and the random variable

$$W$$

follows the beta distribution with distribution function

$$F_W(w) = 1 - (1 - w)^2, 0 < w < 1,$$

Under these assumptions the characteristic function

$$\varphi_U(u)$$

in (3.9.2) has the form

$$\phi_U(u) = (1 - \rho)\frac{\frac{\mu}{iu}\log\left(\frac{\mu}{\mu - iu}\right) - 1}{{}^{iu}\!/_{2\mu}} + \rho\frac{\frac{p\omega\mu}{iu}\log\left(\frac{p\omega\mu}{p\omega\mu - iu}\right) - 1}{{}^{iu}\!/_{2p\omega\mu}}.$$

3.10 Characterizing an Extension of a Random Contraction

It constitutes a direct conclusion of the previous section that the characteristic function

$$\varphi_U(u)$$

of the random contraction of the random variable

$$J = \begin{cases} B, & I = 0 \\ C, & I = 1 \end{cases}$$

via the random variable

$$W,$$

or equivalently the characteristic function of the stochastic multiplicative model

$$U = JW$$

has a complicated form. The present section establishes a characterization for a particular case of that characteristic function.

Theorem 3.10.1 *We suppose that the random variable*

$$W$$

follows the beta distribution with distribution function

$$F_W(w) = w^\tau, \tau > 0, 0 < w < 1$$

and the characteristic function

$$\varphi_U(u)$$

is differentiable then

$$\varphi_U(u) = a \int_0^1 P_N(P_L(uw))w^{a-1}dw,$$

if, and only if,

$$U \overset{d}{=} B, \tag{3.10.1}$$

where

$$\overset{d}{=}$$

denotes equality in distribution and $a = \rho\tau$.

Proof Only the sufficiency condition will be proved since the necessity condition can be proved by reversing the argument. From the previous section and the assumption that the random variable

$$W$$

follows the beta distribution with distribution function

$$F_W(w) = w^\tau, \tau > 0, 0 < w < 1$$

we get that the characteristic function of the random variable

$$U$$

has the form

$$\varphi_U(u) = \tau \int_0^1 [(1-\rho)\varphi_B(uw) + \rho P_N(P_L(\varphi_X(uw)))]w^{\tau-1}dw. \tag{3.10.2}$$

From (2.10.1) it follows that

$$\varphi_U(u) = \varphi_B(u). \tag{3.10.3}$$

If we use (3.10.3) in (3.10.2) we obtain the integral equation

$$\varphi_U(u) = \tau \int_0^1 [(1-\rho)\varphi_U(uw) + \rho P_N(P_L(\varphi_X(uw)))]w^{\tau-1}dw$$

which can be written in the form

$$\varphi_U(u) = \frac{\tau}{u^\tau} \int_0^u [(1-\rho)\varphi_U(uw) + \rho P_N(P_L(\varphi_X(uw)))]w^{\tau-1}dw. \tag{3.10.4}$$

Multiplying both sides of (3.10.4) by u^τ with $u \neq 0$ and differentiating we get the differential equation

$$\tau u^{\tau-1}\varphi_U(u) + u^\tau \frac{d\varphi_U(u)}{du} = \tau[(1-\rho)\varphi_U(u) + \rho P_N(P_L(\varphi_X(u)))]u^{\tau-1} \tag{3.10.5}$$

with the boundary conditions

$$\varphi_U(0) = 1,$$

$$\varphi_X(0) = 1$$

$$P_L(1) = 1,$$

and

$$P_N(1) = 1$$

From (3.10.5) we get that

$$\varphi_U(u) + \frac{u}{\rho\tau}\frac{d\varphi_U(u)}{du} = P_N(P_L(\varphi_X(u))) \tag{3.10.6}$$

for $u \neq 0$. Moreover (3.10.6) implies that

$$\frac{\rho\tau}{u^{\rho\tau}}\int_0^u \left[\varphi_U(w) + \frac{w}{\rho\tau}\frac{d\varphi_U(w)}{dw}\right] w^{\rho\tau-1}dw = \frac{\rho\tau}{u^{\rho\tau}}\int_0^u P_N(P_L(\varphi_X(w)))w^{\rho\tau-1}dw. \tag{3.10.7}$$

It is readily seen (3.10.7) can be written in the form

$$\frac{\rho\tau}{u^{\rho\tau}}\int_0^u \varphi_U(w)w^{\rho\tau-1}dw + \frac{1}{u^{\rho\tau}}\int_0^u w^{\rho\tau-1}d\varphi_U(w)$$
$$= \frac{\rho\tau}{u^{\rho\tau}}\int_0^u P_N(P_L(\varphi_X(w)))w^{\rho\tau-1}dw. \tag{3.10.8}$$

From (3.10.8) we get that

$$\frac{\rho\tau}{u^{\rho\tau}}\int_0^u \varphi_U(w)w^{\rho\tau-1}dw + \varphi_U(w) - \frac{\rho\tau}{u^{\rho\tau}}\int_0^u \varphi_U(w)w^{\rho\tau-1}dw$$
$$= \frac{\rho\tau}{u^{\rho\tau}}\int_0^u P_N(P_L(\varphi_X(w)))w^{\rho\tau-1}dw$$

or equivalently

$$\varphi_U(u) = \frac{\rho\tau}{u^{\rho\tau}}\int_0^u P_N(P_L(\varphi_X(w)))w^{\rho\tau-1}dw. \tag{3.10.9}$$

It is readily seen that (3.10.9) can be written in the form

$$\varphi_U(u) = \rho\tau\int_0^1 P_N(P_L(\phi_X(uw)))w^{\rho\tau-1}dw \tag{3.10.10}$$

From (3.10.10) it follows that

$$\varphi_U(u) = a\int_0^1 P_N(P_L(\varphi_X(uw)))w^{a-1}dw \tag{3.10.11}$$

where

$$a = \rho\tau.$$

The theoretical contribution of the present section consists of establishing a characterization for a particular case of the characteristic function

$$\varphi_U(u)$$

corresponding to the random contraction

$$U = JW.$$

From (3.10.11) it follows that such a characterization represents the distribution corresponding to the random contraction

$$U = JW$$

as a member of the class of α-unimodal distributions [4]. It is very widely adopted that the class of α-unimodal distributions is generally recognized as a strong analytical tool of probability theory with useful applications in various disciplines [2, 3].

Below, we concentrate on the interpretation of the practical contribution of the present section.

We consider the random sum

$$C = X_1 + X_2 + \cdots + X_S$$

with

$$S = L_1 + L_2 + \cdots + L_N.$$

From (3.3.14) we get that the characteristic function of the above random sum has the form

$$\varphi_C(u) = P_N(P_L(\varphi_X(u))).$$

We also consider the random variable

$$W$$

following the beta distribution with distribution function

$$F_W(w) = w^a, a > 0, 0 < w < 1$$

where

$$a = \rho\tau.$$

Moreover, we suppose that the random sum

$$C = X_1 + X_2 + \cdots + X_S$$

is independent of the random variable

$$W.$$

From (3.10.11) it follows that the characteristic function

$$\varphi_U(u)$$

corresponds to the random variable

$$U$$

which can be written in the form

$$U = (X_1 + X_2 + \cdots + X_N)W$$

or equivalently in the form

$$U = CW$$

Hence the random variable

$$U$$

is a random contraction of the random sum

$$C = X_1 + X_2 + \cdots + X_S$$

via the random variable

$$W$$

following the beta distribution with distribution function

$$F_W(w) = w^a, a > 0, 0 < w < 1$$

where

$$a = \rho\tau.$$

Equivalently, the random variable

$$U$$

is a a-unimodal random contraction of the random sum

$$C = X_1 + X_2 + \cdots + X_S$$

where

$$a = \rho\tau.$$

Consequently the practical contribution of the present section consists of the representation of a particular random contraction of the random variable

$$J = \begin{bmatrix} B \ , \ I = 0 \\ \\ C \ , \ I = 1 \end{bmatrix}$$

as a a-unimodal random contraction of the random sum

$$C = X_1 + X_2 + \cdots + X_S.$$

More precisely, such a representation is particularly for the description, investigation, and implementation of control operations arising in global risk governance, cindynics, systemics, informatics, economics, operational research, logistics, marketing, engineering, physics, chemistry, geology, biology, meteorology, and other disciplines of exceptional practical importance.

3.11 Representation of an Extension of a Random Contraction as a Stochastic Discounting Model

The present section is devoted to the representation of the extension of the random contraction of Sect. 3.8 as a stochastic discounting model. It is shown that such a random contraction can be represented as the present value of a random cash flow arising at a future random time. More precisely, the section provides sufficient conditions in order the stochastic discounting model

$$U = Je^{-\delta T}.$$

Where

$$J = \begin{cases} B\ ,\ I = 0 \\ C\ ,\ I = 1, \end{cases}$$

$$\delta$$

a positive real number and

$$T$$

a positive random variable, to be a random contraction. The above stochastic discounting model denotes the present value of a random cash flow under random time and constant force of interest.

Theorem 3.11.1 *Let*

$$N$$

be a discrete random variable with values in the set

$$\boldsymbol{N}_0$$

and let

$$\{L_n, n = 1, 2, ...\}$$

be a sequence of discrete independent random variables with values in the set

$$\boldsymbol{N}_0.$$

We suppose that the random variables of the sequence

$$\{L_n, n = 1, 2, ...\}$$

are equally distributed with the random variable

$$L$$

and we consider the random variable

$$S = L_1 + L_2 + \cdots + L_N.$$

Let

$$\{X_s, s = 1, 2, ...\}$$

be a sequence of positive independent random variables. We suppose that the random variables of the sequence

$$\{X_s, s = 1, 2, ...\},$$

are equally distributed with the random variable

$$X.$$

We set

$$C = X_1 + X_2 + \cdots + X_S$$

and let

$$B$$

be a positive random variable,

$$I$$

a Bernoulli random variable and

$$T$$

a positive random variable. Moreover, we suppose that

$$\delta$$

is a positive real number and we consider the random variable

$$W = e^{-\delta T}$$

with values in the interval

$$(0, 1)$$

If

$$N,$$

$$\{L_n, n = 1, 2, ...\}$$

$$\{X_s, s = 1, 2, ...\},$$

$$B,$$

$$I,$$

and

$$T$$

are independent then the stochastic discounting model

$$U = Je^{-\delta T},$$

or equivalently the stochastic multiplicative model

$$U = JW$$

is a random contraction of the random variable

$$J = \begin{bmatrix} B \ , \ I = 0 \\ \\ C \ , \ I = 1 \end{bmatrix}$$

via the random variable

$$W = e^{-\delta T}$$

Proof From *Theorem* 3.8.1 it follows that the random variable

$$J = \begin{bmatrix} B \ , \ I = 0 \\ \\ C \ , \ I = 1 \end{bmatrix}$$

is independent of the random variable

$$T.$$

Hence the random variable

$$J = \begin{bmatrix} B \ , \ I = 0 \\ \\ C \ , \ I = 1 \end{bmatrix}$$

is also independent of the random variable

$$W = e^{-\delta T}.$$

From the fact that the random variable

$$W = e^{-\delta T}$$

takes values in the interval

$$(0, 1)$$

we conclude that the stochastic multiplicative model

$$U = Je^{-\delta T}$$

or equivalently the stochastic multiplicative

$$U = JW$$

is a random contraction of the random variable

$$J = \begin{bmatrix} B \ , \ I = 0 \\ \\ C \ , \ I = 1 \end{bmatrix}$$

via the random variable

$$W = e^{-\delta T}.$$

An interpretation of the stochastic multiplicative model

$$U = Je^{-\delta T}$$

in the area of continuous discounting is the following.

We suppose that the random variable

$$N$$

denotes the number of different groups of cash flows arising at the random time point

$$T$$

with probability

$$\rho.$$

We also suppose that the random variable

$$L_n, n = 1, 2, ...$$

denotes the number of cash flows consisting the

$$n\text{th}$$

group of cash flows. Hence the random variable

$$S = L_1 + L_2 + \cdots + L_N$$

denotes the total number of cash flows which are incorporated by the

$$N$$

different groups of cash flows at the time point

$$T$$

with probability

$$\rho.$$

Moreover, we suppose that the random variable

$$X_s, s = 1, 2, ...$$

denotes the size of the

$$s\text{th}$$

cash flow at the time point

$$T.$$

Hence the random variable

$$C = X_1 + X_2 + \cdots + X_S.$$

denotes the size of the

$$S = L_1 + L_2 + \cdots + L_N$$

cash flow arising at the time point

$$T$$

with probability

$$\rho.$$

In addition, we suppose that the random variable

$$B$$

denotes the size of a cash flow arising at the time point

$$T$$

with probability

$$1 - \rho.$$

If the positive real number

$$\delta$$

denotes the force of interest the stochastic multiplicative model

$$U = Je^{-\delta T},$$

where

$$J = \begin{cases} B \,, \ I = 0 \\ C \,, \ I = 1, \end{cases}$$

denotes the present value at the time point 0 of the cash flow

$$J$$

arising at the random time

$$T.$$

The theoretical contribution of the present section consists of the interpretation of the formulated random contraction as a stochastic discounting model. The practical contribution of that section consists of making quite clear the vital applicability of such a stochastic model in implementing proactive operations. A significant part of the third chapter of the present work concentrates on the role of these contributions in supporting proactivity in global information risk management operations.

3.12 Characteristic Function of the Representation as a Stochastic Discounting Model of an Extension of a Random Contraction

It is quite obvious that an explicit analytical evaluation of the distribution function

$$F_U(\upsilon)$$

of the random contraction of the random variable

$$J = \begin{bmatrix} B \ , \ I = 0 \\ \\ C \ , \ I = 1 \end{bmatrix}$$

via the random variable

$$W = e^{-\delta T},$$

or equivalently the distribution function of the stochastic multiplicative model

$$U = Je^{-\delta T},$$

is extremely difficult. For this reason the present section is devoted to the evaluation of the corresponding characteristic function

$$\varphi_U(u).$$

Theorem 3.12.1 *Let*

$$F_T(t)$$

be the distribution function of the random variable

$$T$$

and

$$F_W(w) \tag{3.12.1}$$

be the distribution function of the random variable

$$W = e^{-\delta T}$$

then the characteristic function of the random contraction

$$U = JW$$

of the random variable

$$J = \begin{bmatrix} B \ , \ I = 0 \\ C \ , \ I = 1 \end{bmatrix}$$

via the random variable

$$W = e^{-\delta T}$$

or equivalently the characteristic function of the stochastic discounting model

$$U = Je^{-\delta T}$$

has the form

$$\varphi_U(u) = \int_0^1 [(1-\rho)\varphi_B(uw) + \rho P_N(P_L(\varphi_X(uw)))] dF_W(w)$$

or equivalently

$$\varphi_U(u) = \int_0^1 [(1-\rho)\varphi_B(uw) + \rho P_N(P_L(\varphi_X(uw)))] d\left[1 - F_T\left(-\frac{1}{\delta}\log w\right)\right].$$

Proof From (3.9.4) we get that the characteristic function of the random variable

$$J = \begin{bmatrix} B \ , \ I = 0 \\ C \ , \ I = 1 \end{bmatrix}$$

has the form

$$\varphi_J(u) = (1-\rho)\varphi_B(uw) + \rho P_N(P_L(\varphi_X(uw))). \tag{3.12.2}$$

From (3.12.1) it readily follows that the distribution function of the random variable

$$W = e^{-\delta T}$$

is given by

$$F_W(w) = 1 - F_T\left(-\frac{1}{\delta}\log w\right). \tag{3.12.3}$$

Since the random variable

$$J = \begin{cases} B\ , \ I = 0 \\ C\ , \ I = 1 \end{cases}$$

is independent of the random variable

$$W = e^{-\delta T}$$

then from (2.8.1), (3.12.1), (3.12.2) and (3.12.3) it follows that the characteristic function of the random contraction

$$U = JW$$

is given by

$$\varphi_U(u) = \int_0^1 [(1-\rho)\varphi_B(uw) + \rho P_N(P_L(\varphi_X(uw)))]dF_W(w), \tag{3.12.4}$$

or equivalently the characteristic function of the stochastic discounting model

$$U = Je^{-\delta T}$$

is given by

$$\varphi_U(u) = \int_0^1 [(1-\rho)\varphi_B(uw) + \rho P_N(P_L(\varphi_X(uw)))]d\left[1 - F_T\left(-\frac{1}{\delta}\log w\right)\right]$$

Below we consider a particular case of the above characteristic function which is of some practical and theoretical interest.

- We suppose that the random variable

$$N$$

is arbitrarily distributed with probability generating function

$$P_N(z),$$

the random variable

$$L$$

is arbitrarily distributed with probability generating function

$$P_L(z),$$

the random variable

$$X$$

is arbitrary distributed with characteristic function

$$\varphi_X(u),$$

$$N$$

follows the Bernoulli distribution with probability generating function

$$P_I(z) = \rho z + 1 - \rho, 0 < \rho < 1$$

and the random variable

$$T$$

follows the exponential distribution with distribution function

$$F_T(t) = 1 - e^{-\mu t}, \mu > 0, t \geq 0. \tag{3.12.5}$$

From (3.12.3) and (3.12.5) it follows that the distribution function of the random variable

$$W = e^{-\delta T}$$

is given by

$$F_W(w) = w^a, a > 0, 0 < w < 1. \tag{3.12.6}$$

Moreover (3.12.4) and (3.12.6) imply that the characteristic function

$$\varphi_U(u).$$

corresponding to the random contraction

$$U = JW$$

or equivalently corresponding to the stochastic discounting model

$$U = Je^{-\delta T}$$

is given by

$$\varphi_U(u) = a\int_0^1 [(1-\rho)\varphi_B(uw) + \rho P_N(P_L(\varphi_X(uw)))]w^{a-1}dw. \qquad (3.12.7)$$

From (3.12.7) it follows that the characteristic function

$$\varphi_U(u)$$

in (3.12.7) corresponds to a member of the class of α-unimodal random contractions.

A significant part of the third chapter of this work makes use of the theoretical and practical results of the present section, the general acceptance of the exponential distribution as an extremely useful distribution of time, and the consideration of proactivity as a very important structural element of intelligent behavior of systems in order to investigate the role of the proposed interpretation of the extension of the random contraction, as a stochastic discounting model, in global risk governance operations.

3.13 Characterizing the Representation as a Stochastic Discounting Model of an Extension of a Random Contraction

It constitutes a direct conclusion of the previous section that the characteristic function

$$\varphi_U(u)$$

of the random contraction of the random variable

$$J = \begin{bmatrix} B \ , \ I = 0 \\ \\ C \ , \ I = 1 \end{bmatrix}$$

via the random variable

$$W = e^{-\delta T},$$

or equivalently the characteristic function of the stochastic multiplicative model

$$U = Je^{-\delta T},$$

has a very complicated form. The present section establishes a characterization for a particular case of that characteristic function.

Theorem 3.13.1 *We suppose that the random variable*

$$T$$

follows the exponential distribution with distribution function

$$F_T(t) = 1 - e^{-\mu t}, \quad \mu > 0, \quad t \geq 0$$

and the characteristic function

$$\varphi_U(u)$$

is differentiable then

$$\varphi_U(u) = a \int_0^1 P_N(P_L(\varphi_X(uw)))w^{a-1}dw$$

if, and only, if

$$U \stackrel{d}{=} B, \tag{3.13.1}$$

where

$$\stackrel{d}{=}$$

denotes equality in distribution and $a = \frac{\rho\mu}{\delta}$.

Proof Only the sufficiency condition will be proved since the necessity condition can be proved by reversing the argument. From the previous section and the assumption that the random variable

$$T$$

follows the exponential distribution with parameter μ we get that the characteristic function of the random variable

$$U$$

has the form

$$\varphi_U(u) = \vartheta \int_0^1 [(1-\rho)\varphi_B(uw) + \rho P_N(P_L(\varphi_X(uw)))] w^{\vartheta-1} dw, \tag{3.13.2}$$

where

$$\vartheta = \frac{\mu}{\delta}.$$

From (3.13.1) it follows that

$$\varphi_U(u) = \varphi_B(u). \tag{3.13.3}$$

If we use (3.13.3) in (3.13.2) we obtain the integral equation

$$\varphi_U(u) = \vartheta \int_0^1 [(1-\rho)\varphi_U(uw) + \rho P_N(P_L(\varphi_X(uw)))] w^{\vartheta-1} dw$$

which can be written in the form

$$\varphi_U(u) = \frac{\vartheta}{u^{\vartheta}} \int_0^u [(1-\rho)\varphi_U(w) + \rho P_N(P_L(\varphi_X(w)))] w^{\vartheta-1} dw. \tag{3.13.4}$$

Multiplying both sides of (3.13.4) by u^{ϑ} with $u \neq 0$, and differentiating we get the differential equation

$$\vartheta u^{\vartheta-1}\varphi_U(u) + u^{\vartheta}\frac{d\varphi_U(u)}{du} = \vartheta[(1-\rho)\varphi_U(u) + \rho P_N(P_L(\varphi_X(u)))] u^{\vartheta-1} \tag{3.13.5}$$

with the boundary conditions

$$\varphi_U(0) = 1,$$

$$\varphi_X(0) = 1,$$

$$P_L(1) = 1,$$

and

$$P_N(1) = 1.$$

From (3.13.5) we get that

$$\varphi_U(u) + \frac{u}{\rho\vartheta}\frac{d\varphi_U(u)}{du} = P_N(P_L(\varphi_X(u))), \tag{3.13.6}$$

for $u \neq 0$. Moreover (3.13.6) implies that

$$\frac{\rho\vartheta}{u^{\rho\vartheta}}\int_0^u \left[\varphi_U(w) + \frac{w}{\rho\vartheta}\frac{d\varphi_U(w)}{dw}\right] w^{\rho\vartheta-1}dw = \frac{\rho\vartheta}{u^{\rho\vartheta}}\int_0^u P_N(P_L(\varphi_X(w)))w^{\rho\vartheta-1}dw. \tag{3.13.7}$$

It is readily seen that (3.13.7) can be written in the form

$$\frac{\rho\vartheta}{u^{\rho\vartheta}}\int_0^u \varphi_U(w)w^{\rho\vartheta-1}dw + \frac{1}{u^{\rho\vartheta}}\int_0^u w^{\rho\vartheta-1}d\varphi_U(w)$$
$$= \frac{\rho\vartheta}{u^{\rho\vartheta}}\int_0^u P_N(P_L(\varphi_X(w)))w^{\rho\vartheta-1}dw. \tag{3.13.8}$$

From (3.13.8) we get that

$$\frac{\rho\vartheta}{u^{\rho\vartheta}}\int_0^u \varphi_U(w)w^{\rho\vartheta-1}dw + \varphi_U(u) - \frac{\rho\vartheta}{u^{\rho\vartheta}}\int_0^u \varphi_U(w)w^{\rho\vartheta-1}d$$
$$= \frac{\rho\vartheta}{u^{\rho\vartheta}}\int_0^u P_N(P_L(\varphi_X(w)))w^{\rho\vartheta-1}dw$$

or equivalently we get that

$$\varphi_U(u) = \frac{\rho\vartheta}{u^{\rho\vartheta}} \int_0^u P_N(P_L(\varphi_X(w)))w^{\rho\vartheta-1}dw. \tag{3.13.9}$$

It is also readily seen that (3.13.9) can be written in the form

$$\varphi_U(u) = \rho\vartheta \int_0^1 P_N(P_L(\varphi_X(w)))w^{\rho\vartheta-1}dw. \tag{3.13.10}$$

From (3.13.10) it follows that

$$\varphi_U(u) = a \int_0^1 P_N(P_L(\varphi_X(w)))w^{a-1}dw, \tag{3.13.11}$$

where

$$a = \rho\vartheta$$

or equivalently

$$a = \frac{\rho\mu}{\delta}.$$

The theoretical contribution of the present section consists of establishing a characterization for a particular case of the characteristic function

$$\varphi_U(u),$$

corresponding to the random contraction

$$U = Je^{-\delta T}.$$

From (3.13.11) it follows that such a characterization represents the distribution corresponding to the random contraction

$$U = Je^{-\delta T}$$

as a member of the class of α-unimodal distributions [4]. It is very well known that the class of α-unimodal distributions is generally considered as a valuable analytical tool of probability theory with particularly important applications in various disciplines [2, 3].

It is readily understood that there is a very strong theoretical relationship between the characterization established by Sect. 3.10 and the characterization established by Sect. 3.13. It is also readily understood that there is a very significant difference between the practical applicability of these characterizations.

In the following, we concentrate on the clarification of the practical contribution of the present section.

We consider the random sum

$$C = X_1 + X_2 + \cdots + X_S$$

with

$$S = L_1 + L_2 + \cdots + L_N.$$

From (3.3.14) it follows that the characteristic function of the above random sum has the form

$$\varphi_C(u) = P_N(P_L(\varphi_X(w))).$$

We also consider the random variable

$$W$$

following the beta distribution with probability density function

$$f_W(w) = aw^{a-1}, a > 0, 0 < w < 1$$

where

$$a = \frac{\rho\mu}{\delta}.$$

Moreover, we suppose that the random sum

$$C = X_1 + X_2 + \cdots + X_S$$

is independent of the random variable

$$W.$$

From (3.13.11) it follows that the characteristic function

$$\varphi_U(u).$$

corresponds to the random variable

$$U$$

which can be written in the form

$$U = (X_1 + X_2 + \cdots + X_S)W$$

or equivalently in the form

$$U = CW.$$

Hence the random variable

$$U$$

is a random contraction of the random sum

$$C = X_1 + X_2 + \cdots + X_S$$

via the random variable

$$W$$

following the beta distribution with probability density function

$$f_W(w) = w^a, a > 0, 0 < w < 1$$

where

$$a = \frac{\rho\mu}{\delta}.$$

Equivalently, the random variable

$$U$$

is a α-unimodal random contraction of the random sum

$$C = X_1 + X_2 + \cdots + X_S$$

where

$$a = \frac{\rho\mu}{\delta}.$$

In consequence, the practical contribution of the present section consists of the representation of a particular random contraction of the random variable

$$J = \begin{bmatrix} B \, , \, I = 0 \\ \\ C \, , \, I = 1 \end{bmatrix} .$$

as a α-unimodal random contraction of the random sum

$$C = X_1 + X_2 + \cdots + X_S$$

More precisely, such a representation is very useful for the description, investigation, and implementation of proactive control operations arising in risk and crisis management, cindynics, systemics, informatics, economics, operational research, logistics, marketing, engineering, physics, chemistry, geology, biology, meteorology, and other disciplines of exceptional practical importance.

3.14 Conclusions

The contribution of the present chapter includes the following results. More precisely, the contribution of the present chapter consists of the implementation of its twelve theoretical and practical purposes.

- The first result is the formulation of a stochastic multiplicative model. The structural elements of the formulated stochastic multiplicative model are a discrete random variable, a sequence of discrete random variables, a sequence of positive random variables and a positive random variable. The principal components of the formulated stochastic multiplicative model are a sum of a random number of discrete random variables and an extension of the concept of sum of a random number of positive random variables.
- The second result is the establishment of sufficient conditions for the representation of the formulated stochastic multiplicative model as a random contraction.
- The third result is the explicit analytical evaluation of the characteristic function corresponding to such a random contraction.
- The fourth result is the interpretation of the formulated stochastic multiplicative model and the related random contraction as a stochastic discounting model.
- The fifth result is the explicit analytical evaluation of the characteristic function corresponding to such a stochastic discounting model.
- The sixth result is the establishment of an extension of the formulated stochastic multiplicative model by incorporating a positive random variable and a Bernoulli random variable.
- The seventh result is the establishment of sufficient conditions for representing the extension of the formulated stochastic model as a random contraction.
- The explicit analytical evaluation of the characteristic function and the establishment of a characterization of such an extension constitute the eighth and the ninth result of the present chapter respectively.

- The tenth result is the representation of the extension of the formulated stochastic multiplicative model and the related random contraction as a stochastic discounting model.
- The explicit analytical evaluation of the characteristic function and the establishment of a characterization of such a stochastic discounting model constitute the eleventh and the twelfth result of the present chapter respectively.

References

1. P.T. Artikis, C.T. Artikis, Random sums of integral part models in computer systems operations. J. Discret. Math. Sci. Cryptogr. **11**, 209–217 (2008)
2. S. Dharmadhikari, K. Jogdeo, *Unimodality, Convexity, and Applications* (Academic Press Inc., New York, 1988)
3. J. Galambos, I. Simonelli, *Products of Random Variables. Applications to Problems of Physics and to Arithmetical Functions* (Marcel Dekker, New York, 2004)
4. R. Olshen, S. Savage, A generalized unimodality. J. Appl. Probab. **7**, 21–34 (1970)
5. F. Steutel, K. van Harn, *Infinite Divisibility of Probability Distributions on the Real Line* (Marcel Dekker Inc., New York, 2004)
6. H. Tijms, *Stochastic Modeling and Analysis: A Computational Approach*, vol. 13 (Wiley, New York, 1988), p. 1.3

Chapter 4
Interpretations of Random Contractions in Global Risk Governance

Abstract The fourth chapter consists of the incorporation of various well known and particularly significant stochastic models for the precise definition, the thorough investigation and the efficient application of fundamental global risk severity concepts arising in the context of global risk governance. In addition, these stochastic models are readily adopted as particularly significant for the reliable description, the optimal selection and the successful implementation of global or regional risk severity operations in the context of global risk governance. These concepts and operations are generally recognized as very strong tools of global risk governance for the description, investigation and solution of problems with particular practical importance.

Keywords Stochastic Model · Impaired System · Global Risk Severity · Discounting · Global Risk Governance · Unimodality · Stability · Selfdecomposability · Infinite Divisibility

4.1 Introduction

Stochastic models are generally recognized as particularly useful analytical tools for the description, investigation and treatment of significant processes and systems frequently arising in a very wide variety of fundamental disciplines [1, 2]. The present chapter is mainly devoted to the practical interpretations in the context of global risk governance of the stochastic models formulated and investigated by the third chapter of the present work. More precisely, random sums of discrete random variables taking values in the set of positive integers, positive random sums, present values and random contractions of positive random sums, Bernoulli selecting processes based on positive random variables and positive random sums, present values and random contractions of such Bernoulli selecting processes are interpreted as stochastic models suitable for the reliable description, the thorough investigation and the efficient applications of fundamental concepts and proactive operations arising in the context of global risk governance. The main contribution of such interpretations consists of making sufficiently clear that stochastic modeling can be very useful for strongly supporting

P. T. Artikis and C. T. Artikis, *Random Contractions in Global Risk Governance*, Learning and Analytics in Intelligent Systems 27,
https://doi.org/10.1007/978-3-030-95691-2_4

the direct emergence of proactive thinking, planning and acting as extremely indispensable constituent factors of global risk governance. In this work, global risks are clarified as enclosing global geographical extent, strong interindustrial influence and very significant levels of economic and social effect strongly requiring a multistakeholder response. In addition, global risks demand prevalent decision making authorization. This term is employed to make reference, in a general way, to a risk that might have the possibility to impose damage to various systems on a global range [3–5]. Despite the fact that the ongoing and forthcoming risks are of distinct types the consideration of the global risks as a scientific discipline for scientific research is an effective, consistent and significant effort. Many methodological, conceptual and cultural matters arise across the spectrum of global risks. In consequence, some general considerations can be incorporated to various types of global risks and applied to support reliable assessments across the board. There are also quite realistic motives for the recognition of global risks as separate discipline. Observation is not sufficiently frequent and risk reduction is very expensive. The making of decisions on the designation of efforts and resources we need to get qualified deductions. The present chapter can be adopted as a practical contribution to the field of ongoing and forthcoming global risks, useful for an interdisciplinary investigation. There are several particularly significant factors for the established practical results. Researchers mainly concentrating on one of these types of global risks will easily obtain the valuable advantage from learning about other types of global risks. Professionals in disciplines such as insurance, finance and business will also take advantage from a broader aspect [6, 7]. Policy analysts, activists and practitioners involved in the development and implementation of reliable policies likewise stand to earn from learning about the state of the art in investigations concentrating on global risks [8]. The present chapter is also devoted to support further research and awareness and informed public participation about global risks and global risk reduction strategies. The active presence of an interdisciplinary community of experts and practitioners knowledgeable about global risks will significantly improve the obstacles that satisfactory solutions will be obtained and implemented to the difficulties of the current century [9]. Further research activities strongly concentrating on global risk governance are particularly valuable if they are mainly devoted to the formulation, investigation and application of stochastic models suitable for the measurement and treatment of global risks, the development and implementation of methodologies for the accumulation of information persistently required for the predictions of the occurrences of various global risks, the establishment of a stronger interdisciplinary and international risk community containing many academic experts and policy makers responsible for the development and the implementation of global risk governance strategies, and the investigation of global risks and their proactive reduction within the wider and fundamental background of challenges and opportunities for the provision of substantial protection to various human and natural systems [9]. From the fundamental fact that global risks are not enclosed to national borders it directly follows that these risks cannot be effectively managed through the efforts of a single sector. Global risk governance strongly requires an integrative approach incorporating governments, intergovernmental organizations, industry and civil society. It is quite easily understood the great practical contribution of the thorough understanding and the effective management of the arising global risks that have essential effect on human health and safety, the government, the economy and the entire society. Such a practical contribution mainly incorporates the introduction of valuable concepts and principles of global risk governance, the anticipation of notable problems of global risks and the provision of global risk governance proposals for leading decision makers. It constitutes a general adoption that significant advancements in global risk governance are particularly useful in practice if optimal global risk related decisions are to be taken and to substantially increase confidence in vital processes and structures by which crucial decisions are taken. The framework of the international risk governance council presents a forceful structure for the investigation, communication and management of global risks by the systematic incorporation of quite suitable scientific results, global risk perception, social matters and societal principles. Such a framework strives to afford a thorough and combined aspect of global risk governance and hence provide essential support to leading decision makers with making the proper questions and establishing more capable global risk governance strategies for the progressively complex opportunities and global risks frequently arising in many practical situations of the modern world [10, 11]. Stochastic modeling, strategic thinking and global risk governance are considered as very significant scientific disciplines with extremely useful applications in practice. In addition, these scientific disciplines are considered as strongly interrelated [11–14]. The present chapter is mainly devoted to the interpretation of these stochastic models, formulated and investigated by the previous chapter, for the facilitation of strategic thinking and acting in the implementation of principles and goals in the discipline of global risk governance [15].

It is extremely important to mention that the essential contribution of the practical results of the present chapter, in the measurement and proactive treatment of various types of global risks, is mainly based on the fundamental assumption that the occurrences of these global risks can simultaneously impose damages to systems belonging to various countries. In addition, such a fundamental assumption becomes completely acceptable by mentioning that events such as attack or system failure in critical information infrastructure, multiple terrorist attack, extreme weather phenomenon, strong earthquake or seaquake, strong tsunami, major volcanic eruption, geomagnetic storm, rapid and massive infectious diseases, abrupt and major fall in the value of American dollar are generally adopted as causes of global risks whose occurrences can simultaneously affect systems belonging to various countries. It is also very useful to make the clarification that the absence of a generally accepted definition for the concept of global risk and the structural elements of a random contraction constitute the main reasons for the introduction of the assumption that the discrete random variables of the stochastic models, formulated and investigated by the third chapter, of the present work take values in the set of positive integers.

It can be said that discrete random sums are generally recognized as very strong tools of probability theory and particularly useful stochastic models for many fundamental disciplines of extreme practical interest. From the very significant fact that a discrete random variable and a sequence of discrete random variables are readily recognized as the two structural elements of a discrete random sum, it quite easily arises the particularly important necessity for the investigation of the probabilistic behaviour of such a random sum by making very extensive use of the corresponding probability generating function. The theoretical and practical importance of the concept of discrete random sum is mainly based on the powerful result that it is always possible the explicit analytical evaluation of the corresponding probability generating function [16]. More precisely, the explicit analytical evaluation of the probability generating function of a discrete random sum is extremely valuable for the thorough investigation of important theoretical properties of the probability function corresponding to such a random sum. In particular, this kind of thorough investigation is very useful if it mainly concentrates on theoretical properties strongly supporting the extensive applicability of various discrete random sums in modeling, analyzing and managing the behaviour and evolution of systems frequently arising in a very wide variety of practical disciplines. The results of the present chapter provide evidence that global risk governance can be such a practical discipline. Moreover, it is very well known that the concept of discrete random sum quite frequently arises as a constituent factor in the formulation of static or dynamic stochastic models particularly suitable for the reliable description, the thorough investigation and the efficient treatment of extremely significant real world situations strongly related to a wide variety of disciplines with exceptional theoretical and practical importance. In addition, the fourth chapter of the present work makes sufficiently clear that the concept of discrete random sum can substantially contribute to the extension of the practical applicability of various fundamental concepts and proactive operations belonging to the main context of global risk governance.

It constitutes a quite direct adoption that the concept of discrete random sum is properly used by the third chapter of the present work as the most significant factor for the useful introduction of the concept of the positive random sum. More precisely, it is quite easily seen that a concept of discrete random sum is the principal component of the concept of positive random sum. The inclusion of a sequence of positive random variables as a structural element and the inclusion of a discrete random sum as the unique principal component for the formulation of a positive random sum support the particularly important necessity for the investigation of the probabilistic behaviour of such a positive random sum by making very extensive use of the corresponding characteristic function. The theoretical and practical importance of the concept of that positive random sum is mainly based on the powerful result that it is always possible the explicit analytical evaluation of the corresponding characteristic function. In addition, the explicit analytical evaluation of the characteristic function of that positive random sum is extremely useful for the investigation of important theoretical properties of the probability density function of such a positive random sum. In particular, this kind of thorough investigation is valuable if it mainly concentrates on theoretical properties strongly contributing to the extensive applicability of positive random sums in modeling, analyzing and managing the behaviour and evolution of systems frequently arising in a very wide variety of practical disciplines. It can be said that the results of the fourth chapter of the present work provide sufficient evidence for making very clear that global risk governance is easily recognized as a significant member of such a variety of practical disciplines. In consequence, it is quite easily understood the particular importance of the undertaking further research activities mainly concentrating on the establishment extensions of the theoretical results belonging to the third chapter and the practical results belonging to the fourth chapter of the present work.

The main theoretical contribution of the third chapter of the present work consists of the formulation and investigation of stochastic models such as

- *Random Contractions*
- *Discounted Values*
- *Bernoulli Selections*
- *Random Contractions of Bernoulli Selections*
- *Discounted Bernoulli Selections*

by making use of the concept of a positive random sum having as principal component a discrete random sum. The structural elements, the principal component and the mathematical form of such a positive random sum are directly adopted as the main factors supporting the suitability of such a random sum for strongly facilitating the practical applicability of proactive global risk governance operations concentrating on the fundamental concept of global risk severity. Moreover, the fourth chapter of the present work clarifies that the formulation and investigation of random contractions, discounted values, Bernoulli selections, random contractions of Bernoulli selections and discounted Bernoulli selections, by the incorporation of the concept of a positive random sum, can be very useful for the precise description, extensive analysis and efficient implementation of global risk severity reduction operations, global or regional risk severity reduction operations and other proactive operations in the context of global risk governance.

The incorporation of the concept of a positive random sum in the formulation of random contractions, denoting the severity of a global risk after the application of a global risk severity reduction operation, is readily considered as a very significant activity for the proactive treatment of the severity corresponding to a global risk occurring at a future random time point. From the very essential fact that proactive thinking, planning and acting constitute the indispensable advantages of the efficient global risk governance it is easily seen that the formulation, investigation and application of random contractions, based on some positive random sums, strongly contribute to the implementation of the main principles and strategic goals of proactive global risk governance. Moreover, the incorporation of such concept of the positive random sum in the formulation of stochastic models denoting the discounted value of a future cash flow, having the form of that random sum, is easily accepted as a very useful activity for the proactive consideration of the severity of a global risk occurring at a future random time point. From the extremely realistic assumption that proactive thinking, planning and acting are generally considered as fundamental requirements of efficient global risk governance it easily follows that the formulation, investigation and application of discounted positive random sums are very important activities for providing analysts and decision makers with valuable probabilistic information concerning the severity of a global risk. It is also easily accepted that the incorporation of a concept of a positive random sum in the formulation of Bernoulli selecting processes, denoting the severity of a global or regional risk occurrence arising at a future random time point, constitutes a significant activity for the reliable description and the thorough investigation of the impact of such a risk occurrence. It is obvious that the practical contribution in the context of global risk governance of a Bernoulli selecting process, consisting of a positive random variable and some positive random sum, is substantially extended by making the realistic assumption that the positive random variable denotes the severity of a risk with its cause being of low intensity and that positive random sum denotes the severity of the same risk with its cause being of high intensity. In consequence, the formulation of random contractions of Bernoulli selecting processes incorporating a positive random variable and a positive random sum, can be considered very useful for the proactive treatment of the severity of a global or regional risk occurring at a future random time point. More precisely, such random contractions are directly accepted as stochastic models suitable for the description, investigation and implementation of global or regional risk severity reduction operations. It is sufficiently clear that the structural elements, the principal components and the mathematical form of a random contraction corresponding to a Bernoulli selecting process, based on a positive random variable and a positive random sum, strongly support the practical applicability of global or regional risk severity reduction operations. In addition, the formulation of stochastic models denoting the discounted value of a future cash flow, having the form of a Bernoulli selecting process incorporating a positive random variable and a form of positive random sum, constitutes a valuable activity for the proactive consideration of the severity of a global or regional risk occurring at a future random time point. From the fact that proactivity constitutes a fundamental factor of efficient global or regional risk governance it is easily seen that the formulation and application of stochastic models denoting the discounted severity of a global or regional risk, occurring at a future random time point, provide global or regional risk experts with particular useful probabilistic information for proactive thinking, planning and acting.

4.2 Number of Impaired Systems

The number of impaired systems belonging to a number of countries and arising because of the occurrence of a global risk is generally adopted as a very significant factor for the description, development, investigation and application of fundamental concepts and operations particularly useful for proactive thinking, effective deciding and successful acting in the context of global risk governance.

More precisely, the number of impaired systems quite naturally appears as a principal component in concepts such as.

- *Global Risk Severity*
- *Discounted Global Gisk Severity*
- *Global or Regional Risk Severity*

- *Discounted Global or Regional Risk Severity.*

In addition, the number of impaired systems quite naturally appears as a principal component in operations such as.

- *Global Risk Severity Reduction,*
- *Global or Regional Risk Severity Reduction.*

It is easily understood that the number if impaired systems, belonging to a number of countries and arising because of the occurrence of a global risk, is a discrete stochastic model having a discrete random variable and a sequence of discrete random variables as structured elements. It is also easily understood that such a discrete stochastic model has usually the form of a sum consisting of a random number of discrete random variables. In this case, the establishment of sufficient conditions for the representation of such a sum as a random sum constitutes a theoretical contribution of particular practical importance. More clearly, the representation of the number of impaired systems, belonging to a number of countries and arising because of the occurrence of a global risk, as a discrete random sum makes possible the explicit analytical evaluation of the corresponding probability generating function. In other words, it becomes possible the investigation of the probabilistic behaviour of the number of such impaired systems by obtaining and differentiating the corresponding probability generating function. In consequence, it can be said that the explicit analytical representation of the number of systems impaired by the occurrence of a global risk as a discrete random sum may result to a very important support for a particularly useful improvement of the description, development, investigation and implementation of fundamental operations for severity measurement and severity reduction of various global risks. More clearly, the establishment of sufficient conditions for the explicit analytical representation of the number of such impaired systems as a discrete random sum is directly recognized as a very strong reason substantially supporting the establishment of sufficient conditions for the explicit analytical representation of the severity of a global risk as a positive random sum. In consequence, it can be said that the structural elements, the principal component and the explicit analytical form of a positive random sum are quite suitable factors for making extensive use of such a positive random sum in the facilitation of the reliable description, the thorough investigation and efficient implementation of measurement and treatment operations in the context of global risk governance. It is known that the applicability of some positive random sum in global risk governance operations is mainly based on the existence of the explicit analytical form of the corresponding characteristic function.

4.3 Global Risk Severity

The global risk severity or equivalently the severity of a global risk is the economic loss of the number of impaired systems belonging to a number of countries and arising because of the occurrence of a global risk. The global risk severity is generally recognized as an extremely important factor for the description, development, investigation and application of fundamental concepts and operations absolutely necessary for proactive thinking, effective deciding and successful acting in the context of global risk governance.

More precisely, the global risk severity quite naturally appears as a principal component in concepts such as.

- Discounted Global Risk Severity
- Global or Regional Risk Severity
- Discounted Global or Regional Risk Severity.

In addition, global risk severity quite naturally appears as a principal component in operations such as.

- *Global Risk Severity Reduction,*
- *Global or Regional Risk Severity Reduction.*

It is obvious that the global risk severity is a positive stochastic model having a discrete random variable, a sequence of discrete random variables and a sequence of positive random variables as structural elements. Moreover, a sum consisting of a random number of discrete random variables is the principal component of the global risk severity. It is also obvious that the establishment of sufficient condition for the representation of such a positive stochastic model as a random sum constitutes a theoretical contribution of particular practical importance. More clearly, the representation of the global risk severity as a positive random sum makes possible the explicit analytical evaluation of the corresponding characteristic function. In other words, it becomes possible the investigation of the probabilistic behaviour of the global risk severity by obtaining and inverting the corresponding characteristic function. In consequence, it can be said that the stochastic representation of the global risk severity as a positive random sum may result to a very important support for a particularly useful improvement of the description, development, investigation and implementation of fundamental activities for the severity measurement and severity reduction of various global risks.

In the following, it is clarified the practical importance of the theoretical investigation of the characteristic function corresponding to some positive random sum. More precisely, such a theoretical investigation can result to the establishment of sufficient conditions for the presence of unimodality, stability, selfdecomposability, infinite divisibility and other properties in the probability density function of a particular positive random sum. The presence of these properties in a such a probability density function provides decision makers with valuable probabilistic information for the selection and implementation of efficient global risk governance operations concentrating on the measurement and reduction of the severity of various global risks. In addition, the theoretical investigation of the characteristic function of a particular positive random sum can receive substantial facilitation by making extensive use of the fundamental results of the theory of characteristic functions concerning constructions and stochastic derivations of transformations of characteristic functions. Such a facilitation is quite possible because the characteristic function corresponding to a particular positive random sum is very easily recognized as the result of two successive transformations of a characteristic function.

4.4 Discounted Global Risk Severity

The present value, as viewed from the time point 0, of the severity of a global risk occurring at a future random time point is the discounted global risk severity or equivalently the discounted severity of a global risk. It is easily seen that the discounted global risk severity strongly supports proactive thinking, planning and decision making in the context of global risk governance. In addition, the discounted global risk severity quite naturally appears as a principal component in the concept of discounted global or regional risk severity.

It is easily understood that the discounted global risk severity is a positive stochastic model having a discrete random variable, a sequence of discrete random variables, a sequence of positive random variables, a positive random variable and a positive real number as structural elements. Moreover, a sum consisting of a random number of discrete random variables is a principal component of the discounted global risk severity. In addition, such a sum of positive random variables and the discount factor of a single cash flow under random time and constant force of interest are also principal components of the discounted global risk severity. It is also easily understood that the establishment of sufficient conditions for the representation of such a positive stochastic model as a random contraction of a particular positive random sum constitutes a theoretical contribution of particular practical importance. More clearly, the representation of the discounted global risk severity as a random contraction of a particular positive random sum makes possible the explicit analytical evaluation of the corresponding characteristic function. In other words, it becomes possible the investigation of the probabilistic behaviour of the discounted global risk severity by obtaining and inverting the corresponding characteristic function. In consequence, it can be said that the stochastic representation of the discounted global risk severity as a random contraction of a particular positive random sum may result to a very strong support for a particularly useful improvement of the description, development, investigation and implementation of fundamental operations for the proactive treatment of the severity of a global risk.

Below, it is clarified the practical importance of the theoretical investigation of the characteristic function corresponding to the discounted severity of a global risk. More precisely, such an investigation is mainly based on the establishment of sufficient conditions for the representation of the present value of the severity of a global risk as a random contraction. It is given that the severity of a global risk is modeled by a particular positive random sum. Such a representation provides analysts, modelers, decision makers and other experts with valuable probabilistic information for proactive treatment of the severity of a global risk occurring at a future random time point. In particular, random contractions are very suitable stochastic multiplicative models for incorporating unimodality, stability, selfdecomposability, infinite divisibility and other useful analytical properties in the description and solution of significant practical problems. In consequence, it is easily seen that the strong contribution of the discounted global risk severity, having the form of a random contraction of a particular positive random sum, to the efficient and proactive treatment of problems with significant practical importance is mainly based on the explicit analytical evaluation of the characteristic function corresponding to such a random contraction. More precisely, the representation of the discounted global risk severity as a random contraction of some positive random sum strongly supports the presence of unimodality, stability, selfdecomposability, infinite divisibility and other analytical properties which substantially facilitate the description, development, investigation and implementation of various scenarios for the probabilistic behaviour of proactive operations concentrating on the measurement and subsequently reduction of the global risk severity modeled by a particular positive random sum. It is readily adopted that such scenarios completely depend on the explicit analytical evaluation of the characteristic function of the discounted global risk severity being a random contraction.

4.5 Global Risk Severity Reduction

It is quite easily understood that a global risk severity reduction operation is an activity which attacks a global risk by lowering its severity. The severity of a global risk, after the application of a global risk severity reduction operation, is usually the product of a positive random variable with a random variable taking values in the standard unit interval. The positive random variable denotes the severity of the global risk and the random variable taking values in the standard unit interval denotes the impact of the global risk severity reduction operation on the severity of the global risk. It is generally recognized that the stochastic multiplicative model denoting the severity of a global risk, after the application of a global risk severity reduction operation, is a positive stochastic model having a discrete random variable, a sequence of discrete random variables, a sequence of positive random variables and a random variable taking values in the standard unit interval as structural elements. Moreover, a sum consisting of a random number of discrete random variables and such a sum of positive random variables are principal components of the severity of a global risk after the application of a global risk severity reduction operation.

The use of the product of a positive random variable with a random variable taking values in the standard unit interval for the severity of a global risk, after the application of a global risk severity reduction operation, makes theoretically and practically very important the representation of such a product as a random contraction of the severity of a global risk. From a theoretical point of view such a representation very strongly supports the presence of unimodality, stability, selfdecomposability, infinite divisibility and other fundamental analytical properties in the probability density function corresponding to the severity of a global risk after the application of a global risk severity reduction operation. In addition, from a practical point of view these fundamental analytical properties are very useful for providing substantial facilitation to the reliable description, the thorough investigation and the efficient implementation for a wide variety of global risk severity reduction operations in real world situations of particular practical importance.

If the severity of a global risk has the form of some positive random sum then such a sum is the most important principal component of the stochastic multiplicative model denoting the severity of this global risk after the application of a global risk severity reduction operation. From the very important theoretical fact that the characteristic function corresponding to such a positive random sum has an explicit analytical form it easily follows the practical importance of establishing sufficient conditions for the incorporation of random contractions of these random sums in stochastic modeling of global risk severity reduction operations. More precisely, the structural elements, the principal components, the fundamental analytical properties and the characteristic functions corresponding to random contractions of positive random sums are usually considered as particularly suitable tools to provide analysts, modelers, decision makers and other experts with valuable probabilistic information for the reliable description, thorough investigation and efficient implementation of global risk severity reduction operations. In consequence, it can be said that the concept of this positive random sum is recognized as the main factor for the incorporation of random contractions of such random sums as suitable stochastic models for the severity of global risks after the applications of global risk severity reduction operations. It is easily seen that such a recognition is based on the fact that the concept of some positive random sum quite naturally arises as a very stochastic model of the severity of a global risk. In other words, the concept of a particular positive random sum shapes the role of stochastic multiplicative models suitable for the development and implementation of global risk severity reduction operations.

4.6 Global or Regional Risk Severity

The global or regional risk severity is the economic loss of the number of impaired systems belonging to a number of countries and arising because of the occurrence of a global or regional risk. The global or regional risk severity is considered as an important factor for the description, development, investigation and application of fundamental concepts and operations particularly useful for proactive thinking, effective deciding and successful acting in the context of global or regional risk governance. In addition, the global or regional risk severity naturally appears as a principal component in the concept of discounted global or regional risk severity and the operation of global or regional risk severity reduction.

It is easily seen that the global or regional risk severity is a positive stochastic model having a discrete random variable, a sequence of discrete random variables, a sequence of positive random variables, a positive random variable and a Bernoulli random variable as structural elements. Moreover, a sum consisting of a random number of discrete random variables is a principal component of the global or regional risk severity. In addition, such a sum of positive random variables is also a principal component of the global or regional risk severity. It is also easily seen that the establishment of sufficient conditions for the representation of such a positive stochastic model as a positive Bernoulli selecting process constitutes a theoretical contribution of particular practical importance. More clearly, the representation of the global or regional risk severity as a positive Bernoulli selecting process makes possible the explicit analytical evaluation of the corresponding characteristic function. In other words, it becomes possible the investigation of the probabilistic behavior of the global or regional risk severity by obtaining and inverting the corresponding characteristic function. In consequence, it can be said that the stochastic representation of the global or regional risk severity as a positive Bernoulli selecting process may result to a very important support for a particularly useful improvement of the description, development, investigation and implementation of fundamental activities for the severity measurement and severity reduction of various global or regional risks. In the following, it is explained the practical importance of the thorough theoretical investigation of the characteristic function corresponding to a positive Bernoulli selecting process. More precisely, a thorough theoretical investigation of such a characteristic function can result to the establishment of sufficient conditions for the presence of unimodality, stability, selfdecomposability, infinite divisibility and other properties in the probability density function of a positive Bernoulli selecting process. The presence of these properties in such a probability density function provides decision makers with valuable probabilistic information for the selection and implementation of efficient global or regional risk governance operations concentrating on the measurement and reduction of the severity of various global or regional risks. In addition, the theoretical investigation of the characteristic function of a positive Bernoulli selecting process can substantially be facilitated by making extensive use of the fundamental results of the theory of characteristic functions concerning constructions and stochastic derivations of transformations of characteristic functions. Such a facilitation is quite possible because the characteristic function corresponding to a positive Bernoulli selecting process is very easily recognized as the result of three transformations of characteristic functions. It seems to be of significant theoretical importance the clarification that the final result of these three transformations for characteristic functions corresponds to a positive random variable. In consequence, it can be said that the thorough theoretical investigation of the characteristic function corresponding to the formulated positive Bernoulli selecting process is very strongly facilitated by the thorough theoretical investigation of the characteristic function corresponding to some positive random sum. Such a facilitation is particularly useful in considering the existence of a best and most probable scenario for the severity of a global or regional risk.

4.7 Discounted Global or Regional Risk Severity

The present value, as viewed from the time point 0, of the severity of global or regional risk occurring at a future random time point is the discounted global or regional risk severity or equivalently the discounted severity of a global or regional risk. It is easily seen that the discounted global or regional risk severity strongly supports proactive thinking, planning and decision making in the context of global or regional risk governance.

It is easily understood that the discounted global or regional risk severity is a positive stochastic model having a discrete random variable, a sequence of discrete random variables, a sequence of positive random variables, a positive random variable, a Bernoulli random variable, a positive random variable and a positive real number as structural elements. Moreover, a sum consisting of a random number of discrete random variables is a principal component of the discounted global or regional risk severity. In addition, such a sum of positive random variables, a positive Bernoulli selecting process and the discount factor of a single cash flow under random time and constant force of interest are also principal components of the discounted global or regional risk severity. It is also easily understood that the establishment of sufficient conditions for the representation of such a positive stochastic model as a random contraction of a positive Bernoulli selecting process constitutes a theoretical contribution of particular practical importance. More clearly, the representation of the discounted global or regional risk severity as a random contraction of a positive Bernoulli selecting process makes possible the explicit analytical evaluation of the corresponding characteristic function. In other words, it becomes possible the investigation of the probabilistic behaviour of the discounted global or regional risk severity by obtaining and inverting the corresponding characteristic function. In consequence, it can be said that the stochastic representation of the discounted global or regional risk severity as a random contraction of a positive Bernoulli selecting process may result to a very strong support for a particularly useful improvement of the description, investigation and implementation of fundamental operations for the proactive treatment of the severity of a global or regional risk.

In the following, it is clarified the practical importance of the theoretical investigation of the characteristic function corresponding to the discounted severity of a global or regional risk. More precisely, such an investigation is mainly based on the establishment of sufficient conditions for the representation of the present value of the severity of a global or regional risk as a random contraction. It is given that the severity of a global or regional risk is modeled by a positive Bernoulli selecting process. Such a representation provides analysts, modelers, decision makers and other experts with valuable probabilistic information for proactive treatment of the severity of a global or regional risk occurring at a future random time point. In particular, random contractions are quite suitable stochastic multiplicative models for introducing unimodality, stability, selfdecomposability, infinite divisibility and other useful analytical properties in the description and solution of significant practical problems. In consequence, it is obvious that the strong contribution of the discounted global or regional risk severity, having the form of a random contraction of a positive Bernoulli selecting process, to the efficient and proactive treatment of problems with significant practical importance is mainly based on the explicit analytical evaluation of the characteristic function corresponding to such a random contraction. More precisely, the representation of the discounted global or regional risk severity as a random contraction of a positive Bernoulli selecting process strongly supports the presence of unimodality, stability, selfdecomposability, infinite divisibility and other analytical properties which substantially facilitate the description, development, investigation and implementation of various scenarios for the probabilistic behaviour of proactive operations concentrating on the measurement and subsequently reduction of the global or regional risk severity modeled by a positive Bernoulli selecting process. It is readily recognized that such scenarios completely depend on the explicit analytical evaluation of the characteristic function of the discounted global or regional risk severity being a random contraction.

4.8 Global or Regional Risk Severity Reduction

It is easily seen that a global or regional risk severity reduction operation is an activity which attacks a global or regional risk by lowering its severity. The severity of a global or regional risk, after the application of a global or regional risk severity reduction operation, is usually the product of a positive random variable with a random variable taking values in the standard unit interval. The positive random variable denotes the severity of the global or regional risk and the random variable taking values in the standard unit interval denotes the impact of the global or regional risk severity operation on the severity of the global or regional risk. It is readily adopted that the stochastic multiplicative model denoting the severity of a global or regional risk, after the application of a global or regional risk severity reduction operation, is a positive stochastic model having a discrete random variable, a sequence of discrete random variables, a sequence of positive random variables, a positive random variable, a Bernoulli random variable and a random variable taking values in the standard unit interval as structural elements. Moreover, a sum consisting of a random number of discrete random variables is a principal component of the severity of a global or regional risk after the application of a global or regional risk severity reduction operation. In addition, such a sum of positive random variables and a positive Bernoulli selecting process are principal components of the severity of a global or regional risk after the application of a global or regional risk severity reduction operation. It is obvious that the establishment of sufficient conditions for the representation of such a positive stochastic model as a random contraction of a positive Bernoulli selecting process constitutes a theoretical contribution of particular practical importance. More clearly, the representation of the severity of a global or regional risk, after the application of a global or regional risk severity reduction operation, as a random contraction of a positive Bernoulli selecting process makes possible the explicit analytical evaluation of the corresponding characteristic function. In other words, it becomes possible the investigation of the probabilistic behaviour of the severity of a global or regional risk, after the application of a global or regional risk severity reduction operation, by obtaining and inverting the corresponding characteristic function.

If the severity of a global or regional risk has the form of a positive Bernoulli selecting process then such a process is the most important principal component of the stochastic multiplicative model denoting the severity of a global or regional risk after the application of a global or regional risk severity reduction operation. From the fact that the characteristic function corresponding to a positive Bernoulli selecting process has an explicit analytical form it easily follows the practical importance of establishing sufficient conditions for the incorporation of random contractions of Bernoulli selecting processes in stochastic modeling of global or regional risk severity reduction operations. More precisely, the structural elements, the principal components, the fundamental analytical properties and the characteristic functions corresponding to random contractions of positive Bernoulli selecting processes are usually considered as particularly suitable tools to provide analysts, modelers, decision makers and other experts with valuable probabilistic information for the reliable description, thorough investigation and efficient implementation of global or regional risk severity reduction operations. In consequence, it can be said that the concept of positive Bernoulli selecting process is recognized as the main factor for the incorporation of random contractions of such processes as suitable stochastic models for the severity of global or regional risks after the applications of a global or regional risk severity reduction operations. It is easily seen that such a recognition is based on the fact that the concept of positive Bernoulli selecting process quite naturally arises as a very suitable stochastic model of the severity of a global or regional risk. In other words, the concept of positive Bernoulli selecting process shapes the role of stochastic multiplicative models suitable for the development and implementation of global or regional risk severity reduction operations.

4.9 Practical Research Purposes

The present chapter of this work is devoted to the implementation of seven practical purposes. More precisely, the chapter concentrates on the interpretation in the context of global risk governance of the stochastic models formulated and investigated by the previous chapter of this work. It is shown that these formulated stochastic models can be very useful for the description, investigation, selection and application of fundamental concepts and operations of global risk governance. More precisely, the formulated and investigated stochastic models are interpreted as strong analytical tools for very thorough understanding of concepts and operations such as.

- *Number of Impaired Systems,*
- *Global Risk Severity,*
- *Discounted Global Risk Severity,*
- *Global Risk Severity Reduction*

- *Global or Regional Risk Severity,*
- *Discounted Global or Regional Risk Severity,*
- *Global or Regional Risk Severity Reduction.*

In the following, short descriptions of the seven practical purposes of the present chapter are provided.

We suppose that

$$N$$

is a discrete random variable with values in the set

$$\mathbf{N} = \{1, 2, ...\}$$

and

$$\{L_n, n = 1, 2, ...\}$$

is a sequence of discrete random variables with values in the set

$$\mathbf{N}.$$

The first purpose is the interpretation of the random variable

$$S = L_1 + L_2 + ... + L_N$$

as the number of impaired systems, belonging to a random number of countries, because of the occurrence of a global risk at a future random time point.

We suppose that

$$\{X_s, s = 1, 2, ...\}$$

is a sequence of positive random variables. The second purpose is the interpretation of the random variable

$$C = X_1 + X_2 + ... + X_S$$

as the severity of a global risk occurring at a future random time point.

We suppose that

$$T$$

is a positive random variable and

$$\delta$$

is a positive real number. The third purpose is the interpretation of the random variable

$$Y = (X_1 + X_2 + \ldots + X_S)e^{-\delta T}$$

as the present value, as viewed from the time point 0, of the severity of a global risk occurring at a future random time point.

We suppose that

$$W$$

is a random variable with values in the interval

$$(0, 1).$$

The fourth purpose is the interpretation of the random variable

$$Y = (X_1 + X_2 + \ldots + X_S)W$$

as the severity of a global risk, occurring at a future random time point, after the proactive application of a severity reduction operation to the conditions and the cause of such a risk.

We suppose that

$$B$$

is a positive random variable and

$$I$$

is a Bernoulli random variable. The fifth purpose is the interpretation of the random variable

$$J = \begin{cases} B\ , & I = 0 \\ C\ , & I = 1 \end{cases}$$

as the severity of a global or regional risk occurring at a future random time point.

Moreover, the sixth purpose is the interpretation of the random variable

$$U = Je^{-\delta T}$$

as the present value, as viewed from the time point 0, of the severity of a global or regional risk occurring at a future random time point.

In addition, the seventh purpose is the interpretation of the random variable

$$U = JW$$

as the severity of a global or regional risk, occurring at a future random time point, after the proactive application of a risk severity reduction operation to such a risk.

It is also shown that the practical applicability of the seven purposes of the present chapter substantially increases by representing the random variable

$$S = L_1 + L_2 + ... + L_N$$

as a random sum, the random variable

$$C = X_1 + X_2 + ... + X_S$$

as another random sum, the random variable

$$J = \left[\begin{array}{l} B\ ,\ I = 0 \\ \\ C\ ,\ I = 1 \end{array}\right.$$

as a Bernoulli selecting process and the random variables

$$Y = (X_1 + X_2 + ... + X_S)e^{-\delta T},$$

$$Y = (X_1 + X_2 + ... + X_S)W,$$

$$U = Je^{-\delta T}$$

and

$$U = JW$$

as random contractions. These representations are sufficient for evaluating the probability generating function of the random sum

$$S = L_1 + L_2 + ... + L_N,$$

the characteristic function of the random sum

$$C = X_1 + X_2 + ... + X_S,$$

the characteristic function of the random contraction

$$Y = (X_1 + X_2 + ... + X_S)e^{-\delta T},$$

the characteristic function of the random contraction

$$Y = (X_1 + X_2 + ... + X_S)W,$$

the characteristic function of the Bernoulli selecting process

$$J = \begin{bmatrix} B \ , \ I = 0 \\ \\ C \ , \ I = 1 \end{bmatrix},$$

the characteristic function of the random contraction

$$U = Je^{-\delta T}$$

and the characteristic function of the random contraction

$$U = JW,$$

In consequence, the explicit analytical form of the probability generating function corresponding to the random sum

$$S = L_1 + L_2 + ... + L_N,$$

the characteristic function corresponding to the random sum

$$C = X_2 + X_2 + ... + X_S,$$

he characteristic function corresponding to the random contraction

$$Y = (X_1 + X_2 + ... + X_S)e^{-\delta T},$$

the characteristic function corresponding to the random contraction

$$Y = (X_1 + X_2 + ... + X_S)W,$$

the characteristic function corresponding to the Bernoulli selecting process

$$J = \begin{bmatrix} B \ , \ I = 0 \\ \\ C \ , \ I = 1 \end{bmatrix},$$

the characteristic function corresponding to the random contraction

$$U = Je^{-\delta T},$$

and the characteristic function corresponding to the random contraction

$$U = JW$$

support the investigation of the probabilistic behaviour and the extension of the practical applicability of the random variables

$$S = L_1 + L_2 + ... + L_N,$$

$$C = X_1 + X_2 + ... + X_S,$$

$$Y = (X_1 + X_2 + ... + X_S)e^{-\delta T},$$

$$Y = (X_1 + X_2 + ... + X_S)W,$$

$$J = \begin{bmatrix} B \ , \ I = 0 \\ \\ C \ , \ I = 1 \end{bmatrix},$$

$$U = Je^{-\delta T}$$

and

$$U = JW.$$

In particular, the investigation of the probabilistic behaviour of the above random variables mainly concentrates on the presence of the property of unimodality in the probability function corresponding to the discrete random variable

$$S = L_1 + L_2 + ... + L_N$$

and the probability density function for each of the random variables

$$C = X_1 + X_2 + ... + X_S,$$

$$Y = (X_1 + X_2 + ... + X_S)e^{-\delta T},$$

$$Y = (X_1 + X_2 + ... + X_S)W,$$

$$J = \begin{bmatrix} B\ ,\ I = 0 \\ \\ C\ ,\ I = 1 \end{bmatrix},$$

$$U = Je^{-\delta T}$$

and

$$U = JW,$$

The presence of unimodality in the probability function of a discrete random variable or probability density function of a positive random variable implies the existence of a most probable scenario for the factor modeled by such a discrete or positive random variable.

It easily follows that the particular practical importance of the random variables

$$C = X_1 + X_2 + ... + X_S,$$

$$Y = (X_1 + X_2 + ... + X_S)e^{-\delta T},$$

$$Y = (X_1 + X_2 + ... + X_S)W,$$

$$J = \begin{bmatrix} B \ , \ I = 0 \\ \\ C \ , \ I = 1 \end{bmatrix},$$

$$U = Je^{-\delta T}$$

and

$$U = JW$$

substantially increases the interest for investigating the presence of the property of unimodality in the probability density function for each of these random variables. In particular, the interpretation of the random variables

$$C = X_1 + X_2 + ... + X_S,$$

$$Y = (X_1 + X_2 + ... + X_S)e^{-\delta T},$$

$$Y = (X_1 + X_2 + ... + X_S)W,$$

$$J = \begin{bmatrix} B \ , \ I = 0 \\ \\ C \ , \ I = 1 \end{bmatrix},$$

$$U = Je^{-\delta T}$$

and

$$U = JW$$

in the context of global risk governance makes necessary the investigation of the presence of the property of ordinary unimodality in the probability density function for each of these random variables. Such a necessary investigation is supported by the fact that ordinary unimodality means the location of the unique mode at the point 0. In consequence, the presence of ordinary unimodality in the probability density function for each of the random variables

$$C = X_1 + X_2 + ... + X_S,$$

$$Y = (X_1 + X_2 + ... + X_S)e^{-\delta T},$$

$$Y = (X_1 + X_2 + ... + X_S)W,$$

$$J = \begin{bmatrix} B\ ,\ I = 0 \\ \\ C\ ,\ I = 1 \end{bmatrix},$$

$$U = Je^{-\delta T}$$

and

$$U = JW$$

implies the existence of a most probable and best scenario in the interpretation in the context of global risk governance for each of the above random variables.

In general, the establishment of sufficient conditions for introducing the property of ordinary unimodality in the probability density function of a positive random variable is very difficult. Such an establishment is substantially facilitated by embedding such a random variable in the very important class of positive selfdecomposable random variables [17].

4.10 Model of Impaired Systems Number

The present section concentrates on the interpretation of the random variable

$$S = L_1 + L_2 + ... + L_N$$

as a significant factor for the formulation of a stochastic model describing the severity of a global risk.

We suppose that the discrete random variable

$$N$$

denotes the number of countries affected by an occurrence of a global risk. We also suppose that the discrete random variable

$$L_n$$

denotes the number of the impaired systems of the

$$n\text{th}$$

country because of the occurrence of the global risk. Consequently, the discrete random variable

$$S = L_1 + L_2 + ... + L_N$$

denotes the number of the impaired systems of the

$$N$$

countries because of the occurrence of a global risk. It is readily understood that the random variable

$$S = L_1 + L_2 + ... + L_N$$

provides global risk experts with valuable information for the formulation of a stochastic model suitable for the analytical description and the thorough investigation of the severity of a global risk.

If the discrete random variable

$$N$$

has probability generating function

$$P_N(z),$$

the discrete random variables of the sequence

$$\{L_n, n = 1, 2, ...\}$$

are independent and distributed as the random variable

$$L$$

with probability generating function

$$P_L(z)$$

and the random variable

$$N$$

is independent of the sequence

$$\{L_n, n = 1, 2, ...\}$$

then the random variable

$$S = L_1 + L_2 + ... + L_N$$

is a discrete random sum. It is readily shown that the probability generating function of the discrete random sum

$$S = L_1 + L_2 + ... + L_N$$

has the form

$$P_S(z) = P_N(P_L(z)).$$

Since an explicit analytical evaluation of the probability function of the discrete random sum

$$S = L_1 + L_2 + ... + L_N$$

is very difficult then the probability generating function

$$P_S(z) = P_N(P_L(z))$$

constitutes a particularly useful analytical tool for investigating the probability behaviour of the number

$$S = L_1 + L_2 + ... + L_N$$

of the impaired systems of the

$$N$$

countries because of the occurrence of the global risk. The contribution of the present section consists of the establishment of a practical result. Such a result is the interpretation of the random variable

$$S = L_1 + L_2 + ... + L_N$$

as the number of the impaired systems of the

$$N$$

countries because of the occurrence of a global risk. The applicability of such an interpretation is supported by representing the random variable

$$S = L_1 + L_2 + ... + L_N$$

as a random sum. In consequence, the present section supports the undertaking of research activities concentrating on the applicability of discrete random sums in fundamental operations of global risk governance.

It can be said that the combination of the discrete random variable

$$S = L_1 + L_2 + ... + L_N$$

with the sequence of positive random variables

$$\{X_s, s = 1, 2, ...\}$$

for the formulation of a stochastic model with applications in global risk governance is easily recognized as interesting research proposal.

In Sect. 4.1 the random variable

$$S = L_1 + L_2 + ... + L_N$$

is combined with the sequence of positive random variables

$$\{X_s, s = 1, 2, ...\}$$

for the formulation of a stochastic model denoting the severity of a global risk.

If the random variable

$$S = L_1 + L_2 + ... + L_N$$

is a random sum then the difficulty for a direct and an explicit analytical evaluation of the probability function

$$f_S(s)$$

reveals the theoretical and practical importance of the probability generating function

$$P_S(z) = P_N(P_L(z))$$

in investigating the probabilistic behaviour of the number

$$S = L_1 + L_2 + ... + L_N$$

of the impaired systems of the

$$N$$

countries because of the occurrence of a global risk. More precisely, the probability generating function

$$P_S(z) = P_N(P_L(z))$$

provides an indirect and explicit analytical evaluation of the probability function

$$f_S(s).$$

In addition, the probability generating function

$$P_S(z) = P_N(P_L(z))$$

substantially contributes to the investigation of theoretical properties of the probability function

$$f_S(s)$$

which theoretical properties are particularly useful in the investigation and solution of practical problems. The interpretation of the random sum

$$S = L_1 + L_2 + ... + L_N,$$

as the number of the impaired systems of the

$$N$$

countries because of the occurrence of a global risk, becomes extremely useful if the corresponding probability function

$$f_S(s)$$

has the property of unimodality. Equivalently, the random sum

$$S = L_1 + L_2 + ... + L_N$$

can be a significant stochastic model in the context of global risk governance if the corresponding probability function

$$f_S(s)$$

has a unique mode. The existence of a unique mode for the probability function

$$f_S(s)$$

means that a most probable scenario exists for the number

$$S = L_1 + L_2 + ... + L_N$$

of the impaired systems of the

$$N$$

countries because of the occurrence of a global risk. If the probability function

$$f_S(s)$$

has a unique mode at the point

$$s = 0$$

then it is easily understood that the most probable scenario for the number

$$S = L_1 + L_2 + ... + L_N$$

of the impaired systems of the

$$N$$

countries because of the occurrence of a global risk, is also the best scenario. The presence of a unique mode at the point

$$s = 0$$

for the probability function

$$f_S(s)$$

can be established by representing the probability generating function

$$P_S(z) = P_N(P_L(z))$$

as the probability generating function of a renewal distribution [18]. Consequently, the probability generating function

$$P_S(z) = P_N(P_L(z))$$

of the random sum

$$S = L_1 + L_2 + ... + L_N$$

constitutes an analytical tool for investigating the existence of the most probable and best scenario of the number

$$S = L_1 + L_2 + ... + L_N$$

of the impaired systems of the

$$N$$

countries because of the occurrence of a global risk.

Another method for investigating the existence of the most probable and best scenario of the number

$$S = L_1 + L_2 + ... + L_N$$

of the impaired systems of the

$$N$$

countries, because of the occurrence of a global risk, is the representation of the probability generating function

$$P_S(z) = P_N(P_L(z))$$

as the probability generating function of an integral part model [19, 20]. Such a method is particularly useful for developing, investigating and implementing global risk frequency reduction operations [21].

4.11 Model of Global Risk Severity

The present section concentrates on the interpretation of the random variable

$$C = X_1 + X_2 + ... + X_S$$

as a stochastic model for the severity of a global risk.

We suppose that the discrete random variable

$$N$$

denotes the number of countries affected by the occurrence of a global risk and the discrete random variable

$$L_n$$

denotes the number of the impaired systems of the.

$$n\text{th}$$

country because of the occurrence of the global risk. It is easily seen that the discrete random variable

$$S = L_1 + L_2 + ... + L_N$$

denotes the number of the impaired systems of the

$$N$$

countries because of the occurrence of the global risk. We also suppose that the positive random variable

$$X_s$$

denotes the severity of the economic loss of the

$$s\text{th}$$

impaired system. Hence the random variable

$$C = X_1 + X_2 + ... + X_S$$

denotes the severity of the economic losses of the

$$S$$

impaired systems belonging to the

$$N$$

countries because of the occurrence of a global risk. Equivalently, the random variable

$$C = X_1 + X_2 + \dots + X_S$$

denotes the global risk severity.

Moreover, we suppose that the discrete random variable

$$N$$

has probability generating function.

$$P_N(z),$$

the discrete random variables of the sequence

$$\{L_n, n = 1, 2, \dots\}$$

are independent and distributed as the random variable

$$L$$

with probability generating function

$$P_L(z)$$

and the positive random variables of the sequence

$$\{X_s, s = 1, 2, \dots\}$$

are independent and distributed as the random variable

$$X$$

with characteristic function

$$\varphi_X(u).$$

If

$$N.$$

$$\{L_n, n = 1, 2, \dots\}$$

and

$$\{X_s, s = 1, 2, \dots\}$$

are independent then it follows that the random variable

$$S = L_1 + L_2 + ... + L_N$$

is a discrete random sum with probability generating function

$$P_S(z) = P_N(P_L(z))$$

and *Theorem* 3.3.1 implies that the random variable

$$C = X_1 + X_2 + ... + X_S$$

is a positive random sum with characteristic function

$$\varphi_C(u) = P_N(P_L(\varphi_X(u))).$$

Since an explicit analytical evaluation of the distribution function of the positive random sum

$$C = X_1 + X_2 + ... + X_S$$

is impossible then the characteristic function

$$\varphi_C(u) = P_N(P_L(\varphi_X(u)))$$

constitutes a particularly useful analytical tool for investigating the probabilistic behaviour of the severity

$$C = X_1 + X_2 + ... + X_S$$

of a global risk affecting

$$N$$

different countries.

The contribution of the present section consists of the establishment of a practical result. Such a result is the interpretation of the random variable

$$C = X_1 + X_2 + ... + X_S$$

as the severity of a global risk affecting

$$N$$

different countries. The representation of the random variable

$$C = X_1 + X_2 + ... + X_S$$

as a random sum extends the applicability of the established practical result. Consequently, the present section supports the undertaking of research activities concentrating on the applicability of positive random sums in fundamental operations of global risk governance.

In the context of global risk governance, the interpretation of the random variable

$$C = X_1 + X_2 + ... + X_S$$

as the severity of a global risk affecting

$$N$$

different countries means that the global risk severity

$$C = X_1 + X_2 + ... + X_S$$

corresponds to a future random time point

$$T.$$

From the fact that proactivity is generally recognized as an extremely important factor of global risk governance it can be useful to combine the positive random variable

$$C = X_1 + X_2 + ... + X_S$$

with the random variable

$$T$$

for the formulation of a stochastic model denoting the present value, as viewed from the time point 0, of the severity

$$C = X_1 + X_2 + ... + X_S$$

of a global risk occurring at the future random time point

$$T.$$

In Sect. 4.12 the global risk severity

$$C = X_1 + X_2 + ... + X_S$$

and the occurrence time

$$T$$

of a global risk are combined for the formulation of a stochastic model denoting the present value, as viewed from the time point 0, of the severity

$$C = X_1 + X_2 + ... + X_S$$

of a global risk with occurrence time

$$T.$$

The formulation of such a stochastic model is based on the assumption that the force of interest

$$\delta$$

is constant. Moreover, the interpretation of the random variable

$$C = X_1 + X_2 + ... + X_S$$

as the severity of a global risk can be very useful for the formulation, investigation and application of stochastic models suitable for the description and implementation of global risk severity reduction operations.

In Sect. 4.13 the severity

$$C = X_1 + X_2 + ... + X_S$$

of a global risk is combined with a random variable

$$W,$$

taking values in the interval

$$(0, 1),$$

for the formulation of a stochastic multiplicative model suitable for the description and implementation of a global risk severity reduction operation.

It can be said that the representation of the random variable

$$C = X_1 + X_2 + ... + X_S$$

as a random sum constitutes a theoretical contribution with particular practical importance in the description and investigation of the severity of a global risk. It is readily understood that such a contribution is based on the characteristic function

$$\varphi_C(u) = P_N(P_L(\varphi_X(u)))$$

corresponding to the positive random sum

$$C = X_1 + X_2 + ... + X_S.$$

From the fact that the characteristic function

$$\varphi_C(u) = P_N(P_L(\varphi_X(u)))$$

is a transformation of the characteristic function

$$\varphi_X(u)$$

then various results on transformations of characteristic functions can be used for investigating the suitability of the random sum

$$C = X_1 + X_2 + ... + X_S$$

as a stochastic model of the severity of a global risk. More precisely, such results can be valuable for the representation of the transformed characteristic function

$$\varphi_C(u) = P_N(P_L(\varphi_X(u)))$$

as the characteristic function of the random sum

$$C = X_1 + X_2 + ... + X_S$$

with probability density function having the property of unimodality. The use of a unimodal random sum

$$C = X_1 + X_2 + ... + X_S$$

as a stochastic model for the description of the severity of a global risk means the availability of information supporting the existence of a most probable scenario for such a severity. The representation of the transformed characteristic function

$$\varphi_C(u) = P_N(P_L(\varphi_X(u)))$$

as the characteristic function of the random sum

$$C = X_1 + X_2 + ... + X_S,$$

having unimodal probability density function, can be facilitated by embedding

$$\varphi_C(u) = P_N(P_L(\varphi_X(u)))$$

into the class of selfdecomposable characteristic functions [17].

If the random variable

$$C = X_1 + X_2 + ... + X_S$$

is a positive random sum then the difficulty for a direct and explicit analytical evaluation of the probability density function

$$f_C(c)$$

makes quite clear the theoretical and practical importance of the characteristic function

$$\varphi_C(u) = P_N(P_L(\varphi_X(u)))$$

in investigating the probabilistic behaviour of the severity

$$C = X_1 + X_2 + ... + X_S$$

of a global risk affecting

$$N$$

countries. More precisely, the characteristic function

$$\varphi_C(u) = P_N(P_L(\varphi_X(u))),$$

the inversion theorem and the Fast Fourier Transform algorithm make possible an indirect evaluation of the probability density function

$$f_C(c)$$

of the severity

$$C = X_1 + X_2 + ... + X_S$$

of a global risk affecting

$$N$$

countries. It is obvious that the indirect evaluation of the probability density function

$$f_C(c)$$

substantially contributes to investigating the existence of the most probable scenario or the most probable and best scenario for the severity

$$C = X_1 + X_2 + ... + X_S$$

of a global risk affecting

$$N$$

countries. Moreover, the investigation of the existence of the most probable and best scenario for the global risk severity

$$C = X_1 + X_2 + ... + X_S.$$

can be implemented by representing the characteristic function

$$\varphi_C(u) = P_N(P_L(\varphi_X(u)))$$

as the characteristic function of a -unimodal contraction with

$$0 < a \leq 1$$

or as the characteristic function of a ν -unimodal contraction with

$$0 < \nu \leq 1.$$

It is generally adopted that a-unimodal contractions and ν-unimodal contractions are useful for developing, investigating and implementing global risk reduction operations [22, 23].

Another analytical property with significant applications in theory and practice is infinite divisibility. Various stochastic models can only be defined in terms of infinitely divisible distributions. A less basic reason for interest in infinite divisibility is the fact that infinite divisibility of probability distributions in statistics and many areas of applied probability seems to be the rule than the exception. In these cases it may be useful to know that a given probability distribution is infinite divisible, because this implies properties of this probability distribution, such as typical tail behaviour [24]. Consequently, it is necessary to make a comment on the infinite divisibility of the probability distribution of the random sum

$$C = X_1 + X_2 + ... + X_S.$$

It is obvious that the use of the corresponding characteristic function

$$\varphi_C(u) = P_N(P_L(\varphi_X(u)))$$

is suitable for the making of such a comment. We suppose that the discrete random variable

$$N$$

is infinitely divisible then the characteristic function

$$\varphi_C(u) = P_N(P_L(\varphi_X(u)))$$

is a compound Poisson characteristic function. Since a compound Poisson characteristic function is infinitely divisible then the characteristic function

$$\varphi_C(u) = P_N(P_L(\varphi_X(u))),$$

as a compound Poisson characteristic function is infinitely divisible [25]. Compound Poisson characteristic functions play a very important role in the study of infinitely divisible probability distributions. The limit of a sequence of compound Poisson characteristic functions is infinitely divisible. The converse is also true. Every infinitely divisible characteristic function can be written as the limit of a sequence of compound Poisson characteristic functions [25].

Consequently, if the number

$$N$$

of countries, affected by the occurrence of global risk, is an infinitely divisible random variable then the random sum

$$C = X_1 + X_2 + ... + X_S,$$

denoting the global risk severity, is also an infinitely divisible random variable. In consequence, the probability density function

$$f_C(c)$$

of the random sum

$$C = X_1 + X_2 + ... + X_S$$

has heavy tails. In the context of global risk governance, the presence of such tails in the probability density function

$$f_C(c)$$

means that the global risk severity

$$C = X_1 + X_2 + ... + X_S$$

has a significant probability of being small and a significant probability of being big. It is readily understood that a global risk with severity being an infinitely divisible random sum requires a thorough investigation and a proactive treatment.

4.12 Model of Discounted Global Risk Severity

The resent section concentrates on the interpretation of the random variable

$$Y = (X_1 + X_2 + ... + X_S)e^{-\delta T}$$

as a stochastic model for the discounted severity of a global risk occurring at a future random time point.

We suppose that the discrete random variable

$$N$$

denotes the number of countries affected by the occurrence of a global risk at the future random time point

$$T$$

and the discrete random variable

$$L_n$$

denotes the number of the impaired systems of the

$$n\text{th}$$

country because of the occurrence of the global risk. Hence the discrete random variable

$$S = L_1 + L_2 + ... + L_N$$

denotes the number of the impaired systems of the

$$N$$

countries because of the occurrence of the global risk. We also suppose that the positive random variable

$$X_s$$

denotes the severity of the economic loss of the

$$s\text{th}$$

impaired system. Hence the positive random variable

$$C = X_1 + X_2 + ... + X_S$$

denotes the severity of the economic losses of the

$$S$$

impaired systems belonging to the

$$N$$

countries because of the occurrence of the global risk at the future random time point

$$T.$$

Equivalently, the positive random variable

$$C = X_1 + X_2 + ... + X_S$$

denotes the severity of the global risk occurring at the future random time point

$$T.$$

Consequently, if the positive real number

$$\delta$$

denotes the force of interest then the random variable

$$Y = (X_1 + X_2 + ... + X_S)e^{-\delta T}$$

denotes the present value, as viewed from the time point 0, of the severity

$$C = X_1 + X_2 + ... + X_S$$

of a global risk occurring at the future random time point

$$T.$$

Frequently, the random variable

$$Y = (X_1 + X_2 + ... + X_S)e^{-\delta T}$$

is called the discounted severity of a global risk.

Moreover, we suppose that the discrete random variable

$$N$$

has probability generating function

$$P_N(z),$$

the discrete random variables of the sequence

$$\{L_n, n = 1, 2, ...\}$$

are independent and distributed as the random variable

$$L$$

with probability generating function

$$P_L(z),$$

the positive random variables of the sequence

$$\{X_s, s = 1, 2, ...\}$$

are independent and distributed as the random variable

$$X$$

with characterisitc function

$$\varphi_X(u),$$

the positive random variable

$$T$$

has distribution function

$$F_T(t)$$

and we consider the random variable

$$W = e^{-\delta T}$$

taking values in the interval

$$(0, 1).$$

If

$$N.$$

$$\{L_n, n = 1, 2, ...\}.$$

$$\{X_s, s = 1, 2, ...\}$$

and

$$T$$

are independent then *Theorem 3.5.1* implies that the random variable

$$Y = (X_1 + X_2 + ... + X_S)e^{-\delta T}$$

is a random contraction of the random sum

$$C = X_1 + X_2 + ... + X_S$$

via the random variable

$$W = e^{-\delta T}$$

and *Theorem 3.6.1* implies that the characteristic function of the random contraction

$$Y = (X_1 + X_2 + ... + X_S)e^{-\delta T}$$

has the form

$$\varphi_Y(u) = \int_0^1 P_N(P_L(\varphi_X(uw)))d\left[1 - F_T\left(-\frac{1}{\delta}logw\right)\right].$$

In the following, we make clear that the representation of the random variable

$$Y = (X_1 + X_2 + ... + X_S)e^{-\delta T}$$

as a random contraction of the random sum

$$C = X_1 + X_2 + ... + X_S$$

via the random variable

$$W = e^{-\delta T}$$

substantially extends the practical applicability in the context of global risk governance of the interpretation of the random variable

$$Y = (X_1 + X_2 + ... + X_S)e^{-\delta T}$$

as the present value of the severity

$$C = X_1 + X_2 + ... + X_S$$

of a global risk occurring at the future random time point

$$T.$$

More precisely, the representation of the random variable

$$Y = (X_1 + X_2 + ... + X_S)e^{-\delta T}$$

as a random contraction of the random sum

$$C = X_1 + X_2 + ... + X_S$$

via the random variable

$$W = e^{-\delta T}$$

makes possible the explicit analytical evaluation of the characteristic function

$$\varphi_Y(u) = \int_0^1 P_N(P_L(\varphi_X(uw)))d\left[1 - F_T\left(-\frac{1}{\delta}logw\right)\right].$$

Since an explicit analytical and direct evaluation of the probability density function

$$f_Y(y)$$

corresponding to the random contraction

$$Y = (X_1 + X_2 + ... + X_S)e^{-\delta T}$$

is not possible then the investigation of the probabilistic behaviour of the random contraction

$$Y = (X_1 + X_2 + ... + X_S)e^{-\delta T}$$

is implemented by making use of the characteristic function

$$\varphi_Y(u) = \int_0^1 P_N(P_L(\varphi_X(uw)))d\left[1 - F_T\left(-\frac{1}{\delta}logw\right)\right].$$

In particular, we make use of the above characteristic function, the inversion theorem for characteristic functions and the Fast Fourier Transform algorithm in order to obtain the probability density function

$$f_Y(y)$$

of the random contraction

$$Y = (X_1 + X_2 + ... + X_S)e^{-\delta T}$$

denoting the present value of the severity

$$C = X_1 + X_2 + ... + X_S$$

of a global risk occurring at the future random time point

$$T.$$

In other words the representation of the random variable

$$Y = (X_1 + X_2 + ... + X_S)e^{-\delta T}$$

denoting the present value of the severity

$$C = X_1 + X_2 + ... + X_S$$

of a global risk occurring at the future random time point

$$T,$$

as a random contraction of the positive random sum

$$C = X_1 + X_2 + ... + X_S$$

via the random variable

$$W = e^{-\delta T}$$

provides analysts, modelers, decision makers and other experts with valuable probabilistic information for proactive thinking and acting in the context of global risk governance. More precisely, the indirect evaluation of the probability density function

$$f_Y(y)$$

of the random contraction

$$Y = (X_1 + X_2 + ... + X_S)e^{-\delta T}$$

through the characteristic function

$$\varphi_Y(u) = \int_0^1 P_N(P_L(\varphi_X(uw)))d\left[1 - F_T\left(-\frac{1}{\delta}logw\right)\right].$$

gives to the global risk governance experts the extremely important opportunity to thoroughly and proactively investigate the impact of the severity

$$C = X_1 + X_2 + ... + X_S$$

of a global risk occurring at the future random time point

$$T.$$

It is quite obvious that such an investigation strongly supports strategic thinking and acting for the successful treatment of a global risk before its occurrence at the future random time point

$$T.$$

The contribution of the present section can be extended by combing the random contraction

$$Y = (X_1 + X_2 + ... + X_S)e^{-\delta T}$$

denoting the present value of the severity

$$C = X_1 + X_2 + ... + X_S$$

of a global risk occurring at the future random time point

$$T,$$

with the random time interval

$$[0, T).$$

More clearly, such a combination gives the significant advantage to global risk governance experts to make use of the probabilistic information provided by the random contraction

$$Y = (X_1 + X_2 + ... + X_S)e^{-\delta T}$$

for developing, assessing, selecting and implementing fundamental operations for proactive treatment in the random time interval

$$[0, T)$$

of the global risk occurring at the future random time point

$$T.$$

It is of particular practical importance to clarify the interconnection between the existence of a most probable and best scenario for the global risk severity

$$C = X_1 + X_2 + ... + X_S$$

and the existence of a most probable and best scenario for the discounted global risk severity

$$Y = (X_1 + X_2 + ... + X_S)e^{-\delta T}.$$

It is known that the presence of a unique mode at the point

$$c = 0$$

for the probability density function

$$f_C(c)$$

of the random sum

$$C = X_1 + X_2 + ... + X_S$$

implies the presence of a unique mode at the point

$$y = 0$$

for the probability density function

$$f_Y(y)$$

of the random contraction

$$Y = (X_1 + X_2 + ... + X_S)e^{-\delta T}$$

[26]. Equivalently, the existence of a most probable and best scenario for the global risk severity

$$C = X_1 + X_2 + ... + X_S$$

implies the existence of a most probable and best scenario for the discounted global risk severity

$$Y = (X_1 + X_2 + ... + X_S)e^{-\delta T}.$$

Moreover, it is of particular practical importance to provide evidence concerning the role of the random variable

$$T$$

and the random variable

$$W = e^{-\delta T}$$

in the probabilistic behaviour of the random contraction

$$Y = (X_1 + X_2 + ... + X_S)e^{-\delta T}.$$

We suppose that the probability density function

$$f_T(t)$$

of the random variable

$$T$$

satisfies the conditions for making the probability density function

$$f_W(w)$$

of the random variable

$$W = e^{-\delta T}$$

unimodal at the point

$$w = 0.$$

In this case the random contraction

$$Y = (X_1 + X_2 + ... + X_S)e^{-\delta T}$$

has a probability density function

$$f_Y(y)$$

with a unique mode at the point

$$y = 0$$

[26]. Equivalently, a most probable and best scenario exists for the discounted global risk severity

$$Y = (X_1 + X_2 + ... + X_S)e^{-\delta T}$$

based only on the presence of a unique mode at the point

$$w = 0$$

for the probability density function

$$f_W(w).$$

Consequently, it can be said that the role of the random variable

$$T$$

and the role of the random variable

$$W = e^{-\delta T}$$

are important for thinking, deciding and acting proactively in the context of global risk governance.

For example, if the random variable

$$T$$

follows the exponential distribution with parameter

$$\mu$$

Satisfying

$$\mu < \delta$$

then the discounted global risk severity

$$Y = (X_1 + X_2 + ... + X_S)e^{-\delta T}$$

is a a -unimodal contraction with probability density function

$$f_Y(y)$$

having a unique mode at the point

$$y = 0,$$

where

$$a = \mu/\delta.$$

4.13 Model of Global Risk Severity Reduction

The present section concentrates on the interpretation of the random variable

$$Y = (X_1 + X_2 + ... + X_S)W$$

as a stochastic model for the severity of a global risk after the application of a global risk severity reduction operation.

We suppose that the discrete random variable

$$N$$

denotes the number of countries affected by the occurrence of a global risk and the discrete random variable

$$L_n$$

denotes the number of the impaired systems of the

$$n\text{th}$$

country because of the occurrence of the global risk. It is obvious that the discrete random variable

$$S = L_1 + L_2 + ... + L_N$$

denotes the number of the impaired systems of the

$$N$$

countries because of the occurrence of the global risk. We also suppose that the positive random variable

$$X_s$$

denotes the severity of the economic loss of the

$$s\text{th}$$

impaired system. Hence the positive random variable

$$C = X_1 + X_2 + ... + X_S$$

denotes the severity of the global risk. In addition, we suppose that the random variable

$$W,$$

with values in the interval

$$(0, 1),$$

denotes the impact of applying a global risk severity reduction operation to the conditions and the cause of the global risk. Under the above assumptions, the random variable

$$Y = (X_1 + X_2 + ... + X_S)W$$

can be interpreted as the severity of the global risk after the application of a global risk severity reduction operation.

Moreover, we suppose that the discrete random variable

$$N$$

has probability generating function

$$P_N(z),$$

the discrete random variables of the sequence

$$\{L_n, n = 1, 2, ...\}$$

are independent and distributed as the random variable

$$L$$

with probability generating function

$$P_L(z),$$

the positive random variables of the sequence

$$\{X_s, s = 1, 2, ...\}$$

are independent and distributed as the random variable

$$X$$

with characterisitc function

$$\varphi_X(u)$$

and the random variable

$$W$$

has distribution function

$$F_W(w).$$

If

$$N,$$

$$\{L_n, n = 1, 2, ...\},$$

$$\{X_s, s = 1, 2, ...\}$$

and

$$W$$

are independent then *Theorem3.3.1* implies that the random variable

$$Y = (X_1 + X_2 + ... + X_S)W$$

is a random contraction of the positive random sum

$$C = X_1 + X_2 + ... + X_S$$

via the random variable

$$W$$

and *Theorem* 3.4.1 implies that the characteristic function of the random contraction

$$Y = (X_1 + X_2 + ... + X_S)W$$

has the form

$$\varphi_Y(u) = \int_0^1 P_N(P_L(\varphi_X(uw)))dF_W(w).$$

In the following, we make clear that the representation of the random variable

$$Y = (X_1 + X_2 + ... + X_S)W$$

as a random contraction of the random sum

$$C = X_1 + X_2 + ... + X_S$$

via the random variable

$$W$$

substantially extends the practical applicability in the context of global risk governance of the interpretation of the random variable

$$Y = (X_1 + X_2 + ... + X_S)W$$

as the severity of the global risk after the application of a global risk severity reduction operation to the conditions and the cause of such a risk. More precisely, the representation of the random variable

$$Y = (X_1 + X_2 + ... + X_S)W$$

as a random contraction of the random sum

$$C = X_1 + X_2 + ... + X_S$$

via the random variable

$$W$$

makes possible the explicit analytical evaluation of the characteristic function

$$\varphi_Y(u) = \int_0^1 P_N(P_L(\varphi_X(uw)))dF_W(w).$$

Since an explicit analytical and direct evaluation of the probability density function

$$f_Y(y)$$

corresponding to the random contraction

$$Y = (X_1 + X_2 + ... + X_S)W$$

is not possible then the investigation of the probabilistic behaviour of the random contraction

$$Y = (X_1 + X_2 + ... + X_S)W$$

is implemented by making use of the characteristic function

$$\varphi_Y(u) = \int_0^1 P_N(P_L(\varphi_X(uw)))dF_W(w).$$

In particular, we make use of the above characteristic function, the inversion theorem for characteristic functions and the Fast Fourier Transform algorithm in order to obtain the probability density function

$$f_Y(y)$$

of the random contraction

$$Y = (X_1 + X_2 + ... + X_S)W$$

denoting the severity of the global risk after the application of a global risk severity reduction operation to the conditions and the cause of such a risk.

In other words, the representation of the random variable

$$Y = (X_1 + X_2 + ... + X_S)W,$$

denoting the severity of the global risk after the application of a global risk severity reduction operation to the conditions and the cause of such a risk, as a random contraction of the random sum

$$C = X_1 + X_2 + ... + X_S$$

via the random variable

$$W$$

provides analysts, modelers, decision makers and other experts with valuable probabilistic information for proactive treatment of global risks. More precisely, the indirect evaluation of the probability density function

$$f_Y(y)$$

of the random contraction

$$Y = (X_1 + X_2 + ... + X_S)W$$

through the characteristic function

$$\varphi_Y(u) = \int_0^1 P_N(P_L(\varphi_X(uw)))dF_W(w)$$

gives to the global risk governance experts the extremely important opportunity to proactively reduce the severity

$$C = X_1 + X_2 + ... + X_S$$

of a global risk before its occurrence.

The contribution of the present section can be extended by combining the random contraction

$$Y = (X_1 + X_2 + ... + X_S)W,$$

denoting the severity of a global risk after the application of global risk severity reduction operation to such a risk, with the random time interval

$$[0, T),$$

where

$$T$$

is a future random time point of the occurrence of the global risk. More precisely, such a combination gives the significant advantage to global risk governance experts to make use of the probabilistic information provided by the random contraction

$$Y = (X_1 + X_2 + ... + X_S)W$$

for developing, assessing, selecting and implementing global risk severity reduction operations for proactive treatment in the random time interval

$$[0, T)$$

of the global risk occurring at the future random time point

$$T.$$

It is of particular practical importance to clarify the interconnection between the existence of a most probable and best scenario for the global risk severity

$$C = X_1 + X_2 + ... + X_S$$

and a most probable and best scenario for the global risk severity

$$Y = (X_1 + X_2 + ... + X_S)W$$

after the application of a global risk severity reduction operation. It is known that the presence of a unique mode at the point

$$c = 0$$

for the probability density function

$$f_C(c)$$

of the random sum

$$C = X_1 + X_2 + ... + X_S$$

implies the presence of a unique mode at the point

$$y = 0$$

for the probability density function

$$f_Y(y)$$

of the random contraction

$$Y = (X_1 + X_2 + ... + X_S)W$$

[26]. Equivalently, the existence of a most probable and best scenario for the global risk severity

$$C = X_1 + X_2 + ... + X_S$$

implies the existence of a most probable and best scenario for the global risk severity

$$Y = (X_1 + X_2 + ... + X_S)W$$

after the application of a global risk severity reduction operation.

Moreover, it is of particular practical importance to provide evidence concerning the role of the random variable

$$W$$

in the probabilistic behaviour of the random contraction

$$Y = (X_1 + X_2 + ... + X_S)W.$$

It is known that the presence of a unique mode at the point

$$w = 0$$

for the probability density function

$$f_W(w)$$

of the random variable

$$W$$

implies the presence of a unique mode at the point

$$y = 0$$

for the probability density function

$$f_Y(y)$$

of the random contraction

$$Y = (X_1 + X_2 + \ldots + X_S)W$$

[26]. Equivalently, a most probable and best scenario for the global risk severity

$$Y = (X_1 + X_2 + \ldots + X_S)W,$$

after the application of a global risk severity reduction operation, based only on the presence of a unique mode at the point

$$w = 0$$

for the probability density function

$$f_W(w)$$

of the random variable

$$W.$$

Consequently, it can be said that the role of the random variable

$$W$$

is important in selecting and implementing a global risk severity reduction operation. More clearly, the selection and implementation of an efficient global risk severity reduction operation can substantially contribute to a proactive and successful treatment of a global risk. Below, we consider two interesting cases for the probability density function

$$f_W(w)$$

of the random variable

$$W.$$

We suppose that the random variable

$$W$$

follows the beta distribution with probability density function

$$f_W(w) = aw^{a-1}, a > 0, 0 < w < 1,$$

which has a unique mode at the point

$$w = 0$$

under the assumption that

$$0 < a < 1.$$

It is known that under the above assumption the random variable

$$Y = (X_1 + X_2 + ... + X_S)W$$

is a α-unimodal random contraction with probability density function

$$f_Y(y)$$

having a unique mode at the point

$$y = 0$$

[26]. Consequently, if the random variable

$$W,$$

following the beta distribution with probability density function

$$f_W(w) = aw^{a-1}, a > 0, 0 < w < 1,$$

denotes the impact of a global risk severity reduction operation applied to the cause and the conditions of a global risk then a most probable and best scenario exists for the severity

$$Y = (X_1 + X_2 + ... + X_S)W$$

of such a global risk.

Moreover, we suppose that the random variable

$$W$$

follows the beta distribution with probability density function

$$f_W(w) = \nu(1 - w)^{\nu-1}, \nu > 0, 0 < w < 1,$$

which has a unique mode at the point

$$w = 0$$

under the assumption that

$$\nu \geq 1.$$

It is known that under the above assumption the random variable

$$Y = (X_1 + X_2 + ... + X_S)W$$

is a υ -unimodal random contraction with probability density function

$$f_Y(y)$$

having a unique mode at the point

$$y = 0$$

[26]. Consequently, if the random variable

$$W,$$

following the beta distribution with probability density function

$$f_W(w) = \nu(1 - w)^{\nu-1}, \nu \geq 1, 0 < w < 1,$$

denotes the impact of a global risk severity reduction operation applied to the cause and the conditions of a global risk then a most probable and best scenario exists for the severity

$$Y = (X_1 + X_2 + ... + X_S)W$$

of such a global risk.

4.14 Model of Global or Regional Risk

The present section concentrates on the interpretation of the random variable

$$J = \begin{bmatrix} B\ ,\ I = 0 \\ \\ C\ ,\ I = 1 \end{bmatrix}$$

as a stochastic model for the severity of a global or regional risk.

We suppose that a risk occurs. The risk is global with probability

$$\rho$$

and the risk is regional with probability

$$1 - \rho.$$

We also suppose that the discrete random variable

$$N,$$

denotes the number of countries affected by the occurrence of the global risk and the discrete random variable

$$L_n$$

denotes the number of the impaired systems of the

$$n\text{th}$$

country because of the occurrence of the global risk. Hence the discrete random variable

$$S = L_1 + L_2 + ... + L_N$$

denotes the number of the impaired systems of the

$$N$$

countries because of the occurrence of the global risk. In addition, we suppose that the positive random variable

$$X_s$$

denotes the severity of the economic loss of the

$$s\text{th}$$

impaired system. Hence the positive random variable

$$C = X_1 + X_2 + ... + X_S$$

denotes the severity of the global risk occurring with probability

$$\rho.$$

Moreover, we suppose that the positive random variable

$$B$$

denotes the severity of the regional risk occurring with probability

$$1 - \rho.$$

If

$$I$$

is a Bernoulli random variable with probability function

$$P(I = 0) = 1 - \rho,$$

$$P(I = 1) = \rho$$

then the random variable

$$J = \begin{cases} B \ , \ I = 0 \\ C \ , \ I = 1 \end{cases}$$

denotes the severity of the global or regional risk.

Furthermore, we suppose that the discrete random variable

$$N$$

has probability generating function

$$P_N(z),$$

the discrete random variables of the sequence

$$\{L_n, n = 1, 2, ...\}$$

are independent and distributed as the random variable

$$L$$

with probability generating function

$$P_L(z)$$

the positive random variables of the sequence

$$\{X_s, s = 1, 2, ...\}$$

are independent and distributed as the random variable

$$X$$

with characteristic function

$$\varphi_X(u),$$

the positive random variable

$$B$$

has characteristic function

$$\varphi_B(u)$$

and the Bernoulli random variable

$$I$$

has probability generating function

$$P_I(z) = 1 - \rho + \rho z.$$

If

$$N,$$

$$\{L_n n = 1, 2, ...\}$$

$$\{X_s, s = 1, 2, ...\},$$

$$B,$$

and

$$I$$

are independent then *Theorem 3.8.1* implies that the characteristic function of the random variable

$$J = \begin{bmatrix} B \ , \ I = 0 \\ \\ C \ , \ I = 1 \end{bmatrix}$$

has the form

$$\varphi_J(u) = (1 - \rho)\varphi_B(u) + \rho P_N(P_L(\varphi_X(u))).$$

Since an explicit analytical and direct evaluation of the probability density function

$$f_J(j)$$

corresponding to the random variable

$$J = \begin{bmatrix} B \ , \ I = 0 \\ \\ C \ , \ I = 1 \end{bmatrix}$$

is not possible then the investigation of the probabilistic behaviour of the random variable

$$J = \begin{bmatrix} B \ , \ I = 0 \\ \\ C \ , \ I = 1 \end{bmatrix}$$

is implemented by making use of the characteristic function

$$\varphi_J(u) = (1 - \rho)\varphi_B(u) + \rho P_N(P_L(\varphi_X(u))).$$

In particular, we make use of the above characteristic function, the inversion theorem for characteristic functions and the Fast Fourier Transform algorithm in order to obtain the probability density function

$$f_J(j)$$

of the random variable

$$J = \begin{bmatrix} B \ , \ I = 0 \\ \\ C \ , \ I = 1 \end{bmatrix}$$

denoting the severity of a global or regional risk. The indirect evaluation of the probability density function

$$f_J(j)$$

of the random variable

$$J = \begin{bmatrix} B \ , \ I = 0 \\ \\ C \ , \ I = 1 \end{bmatrix}$$

through the characteristic function

$$\varphi_J(u) = (1 - \rho)\varphi_B(u) + \rho P_N(P_L(\varphi_X(u))).$$

gives to the global risk governance experts the extremely important opportunity to thoroughly and proactively investigate the impact of the severity

$$J = \begin{bmatrix} B \ , \ I = 0 \\ \\ C \ , \ I = 1 \end{bmatrix}$$

of a global or regional risk. It is of particular practical importance to clarify the interconnection between the existence of a most probable and best scenario for the severity

$$B$$

of a regional risk, the existence of a most probable and best scenario for the severity

$$C = X_1 + X_2 + ... + X_S$$

of a global risk and the existence of a most probable and best scenario for the severity

$$J = \begin{bmatrix} B \ , \ I = 0 \\ \\ C \ , \ I = 1 \end{bmatrix}$$

of a global or regional risk. It is known that the presence of unique mode at the point

$$\beta = 0$$

for the probability density function

$$f_B(\beta)$$

of the positive random variable

$$B$$

and the presence of unique mode at the point

$$c = 0$$

for the probability density function

$$f_C(c)$$

of the positive random sum

$$C = X_1 + X_2 + ... + X_S$$

implies the presence of unique mode at the point

$$j = 0$$

for the probability density function

$$f_J(j)$$

of the random variable

$$J = \begin{bmatrix} B\ ,\ I = 0 \\ \\ C\ ,\ I = 1 \end{bmatrix}$$

[26]. In consequence, the existence of a most probable and best scenario for the severity

$$B$$

of a regional risk and the existence of a most probable and best scenario for the severity

$$C = X_1 + X_2 + ... + X_S$$

of a global risk imply the existence of a most probable and best scenario for the severity

$$J = \begin{cases} B \ , \ I = 0 \\ C \ , \ I = 1 \end{cases}$$

of a global or regional risk.

The contribution of the present section consists of the interpretation of the random variable

$$J = \begin{cases} B \ , \ I = 0 \\ C \ , \ I = 1 \end{cases}$$

as the severity of a global or regional risk. It easily follows that the contribution of the present section is substantially extended by making the assumption that the random variable

$$B$$

denotes the severity of a risk with its cause being of low intensity and the random variable

$$C = X_1 + X_2 + ... + X_S$$

denotes the severity of the same risk with its cause being of high intensity.

In the context of global risk governance, the interpretation of the random variable

$$J = \begin{cases} B \ , \ I = 0 \\ C \ , \ I = 1 \end{cases}$$

as the severity of a global or regional risk means that the severity

$$J = \begin{cases} B \ , \ I = 0 \\ C \ , \ I = 1 \end{cases}$$

corresponds to a future random time point

$$T.$$

From the fact that proactivity is generally recognized as an extremely important factor of global risk governance it can be useful to combine the random variable

$$J = \begin{cases} B \,, & I = 0 \\ C \,, & I = 1 \end{cases}$$

with the random variable

$$T$$

for the formulation of a stochastic model denoting the present value, as viewed from the time point 0, of the severity

$$J = \begin{cases} B \,, & I = 0 \\ C \,, & I = 1 \end{cases}$$

of a global or regional risk occurring at the future time point

$$T.$$

In Sect. 4.15 the severity

$$J = \begin{cases} B \,, & I = 0 \\ C \,, & I = 1 \end{cases}$$

and the occurrence time

$$T$$

of a global or regional risk are combined for the formulation of a stochastic model denoting the present value, as viewed from the time point 0, of the severity

$$J = \begin{cases} B \,, & I = 0 \\ C \,, & I = 1 \end{cases}$$

of a global or regional risk with occurrence time

$$T.$$

The formulation of such a stochastic model is based on the assumption that the force of interest

$$\delta$$

is constant. Moreover, the interpretation of the random variable

$$J = \begin{cases} B \ , \ I = 0 \\ C \ , \ I = 1 \end{cases}$$

as the severity of a global or regional risk can be very useful for the formulation, investigation and application of stochastic models suitable for the description and implementation of global or regional risk severity reduction operations.

In Sect. 4.16 the severity

$$J = \begin{cases} B \ , \ I = 0 \\ C \ , \ I = 1 \end{cases}$$

of a global or regional risk is combined with the random variable

$$W$$

taking values in the interval

$$(0, 1)$$

for the formulation of a stochastic multiplicative model suitable for the description and implementation of a global or regional risk severity reduction operation.

4.15 Model of Discounted Global or Regional Risk Severity

The present section concentrates on the interpretation of the random variable

$$U = Je^{-\delta T}$$

as a stochastic model for the discounted severity of a global or regional risk occurring at a future random time point.

We suppose that a risk occurs at the future random time point

$$T.$$

The risk is global with probability

$$\rho$$

and the risk is regional with probability

$$1 - \rho.$$

We also suppose that the discrete random variable

$$N$$

denotes the number of countries affected by the occurrence of the global risk and the discrete random variable

$$L_n$$

denotes the number of the impaired systems of the

$$n\text{th}$$

country because of the occurrence of the global risk. Hence the discrete random variable

$$S = L_1 + L_2 + \ldots + L_N.$$

denotes the number of the impaired systems of the

$$N$$

countries because of the occurrence of the global risk. In addition, we suppose that the positive random variable

$$X_s$$

denotes the severity of the economic loss of the

$$s\text{th}$$

impaired system. Hence the positive random variable

$$C = X_1 + X_2 + \ldots + X_S$$

denotes the severity of the global risk occurring at the future random time point

$$T$$

with probability

$$\rho.$$

Moreover, we suppose that the positive random variable

$$B$$

denotes the severity of the regional risk occurring at the future random time point

$$T$$

with probability

$$1 - \rho.$$

If

$$I$$

is a Bernoulli random variable with probability function

$$P(I = 0) = 1 - \rho,$$

$$P(I = 1) = \rho$$

then the random variable

$$J = \begin{cases} B\ , I = 0 \\ C\ , I = 1 \end{cases}$$

denotes the severity of the global or regional risk at the future random time point

$$T.$$

In consequence, the random variable

$$U = Je^{-\delta T}$$

denotes the present value, as viewed from the time point 0, of the severity

$$J = \begin{cases} B \ , \ I = 0 \\ C \ , \ I = 1 \end{cases}$$

of the global or regional risk occurring at the future random time point

$$T,$$

where the positive real number

$$\delta$$

denotes the force of interest in continuous discounting.

Furthermore, we suppose that the discrete random variable

$$N$$

has probability generating function

$$P_N(z),$$

the discrete random variables of the sequence

$$\{L_n, n = 1, 2, ...\}$$

are independent and distributed as the random variable

$$L$$

with probability generating function

$$P_L(z),$$

the positive random variables of the sequence

$$\{X_s, s = 1, 2, ...\}$$

are independent and distributed as the random variable

$$X$$

with characteristic function

$$\varphi_X(u).$$

the positive random variable

$$B$$

has characteristic function

$$\varphi_B(u).$$

the Bernoulli random variable

$$I$$

has probability generating function

$$P_I(z) = 1 - \rho + \rho z,$$

the positive random variable

$$T$$

has distribution function

$$F_T(t)$$

and we consider the random variable

$$W = e^{-\delta T}$$

taking values in the interval

$$(0, 1).$$

If

$$N.$$

$$\{L_n, n = 1, 2, ...\},$$

$$\{X_s, s = 1, 2, ...\},$$

$$B,$$

$$I,$$

and

$$T$$

are independent then *Theorem 3.11.1* implies that the random variable

$$U = Je^{-\delta T}$$

is a random contraction of the random variable

$$J = \begin{bmatrix} B\ , I = 0 \\ \\ C\ , I = 1 \end{bmatrix}$$

via the random variable

$$W = e^{-\delta T}$$

and *Theorem 3.12.1* implies that the characteristic function of the random contraction

$$U = Je^{-\delta T}$$

has the form

$$\varphi_U(u) = \int_0^1 [(1-\rho)\varphi_B(uw) + \rho P_N(P_L(\varphi_X(uw)))]d\left[1 - F_T\left(-\frac{1}{\delta}logw\right)\right].$$

In the following we make clear that the representation of the random variable

$$U = Je^{-\delta T}$$

as a random contraction of the random variable

$$J = \begin{bmatrix} B\ , I = 0 \\ \\ C\ , I = 1 \end{bmatrix}$$

via the random variable

$$W = e^{-\delta T}$$

substantially extends the practical applicability in the context of global risk governance of the interpretation of the random variable

$$U = Je^{-\delta T}$$

as the present value of the severity

$$J = \begin{cases} B \ , \ I = 0 \\ C \ , \ I = 1 \end{cases}$$

of a global or regional risk occurring at the future time point

$$T.$$

More precisely, the representation of the random variable

$$U = Je^{-\delta T}$$

as a random contraction of the random variable

$$J = \begin{cases} B \ , \ I = 0 \\ C \ , \ I = 1 \end{cases}$$

via the random variable

$$W = e^{-\delta T}$$

makes possible the explicit analytical evaluation of the characteristic function

$$\varphi_U(u) = \int_0^1 [(1-\rho)\varphi_B(uw) + \rho P_N(P_L(\varphi_X(uw)))]d\left[1 - F_T\left(-\frac{1}{\delta}logw\right)\right].$$

Since an explicit analytical and direct evaluation of the probability density function

$$f_U(\upsilon)$$

corresponding to the random contraction

$$U = Je^{-\delta T}$$

is not possible then the investigation of the probabilistic behaviour of the random contraction

$$U = Je^{-\delta T}$$

is implemented by making use of the characteristic function

$$\varphi_U(u) = \int_0^1 [(1-\rho)\varphi_B(uw) + \rho P_N(P_L(\varphi_X(uw)))]d\left[1 - F_T\left(-\frac{1}{\delta}logw\right)\right].$$

In particular, we make use of the above characteristic function, the inversion theorem for characteristic functions and the Fast Fourier Transform algorithm in order to obtain the probability density function

$$f_U(\upsilon)$$

of the random contraction

$$U = Je^{-\delta T}$$

denoting the present value of the severity

$$J = \begin{bmatrix} B \ , \ I = 0 \\ C \ , \ I = 1 \end{bmatrix}$$

of a global or regional risk occurring at the future time point

$$T.$$

In other words, the representation of the random variable

$$U = Je^{-\delta T},$$

denoting the present value of the severity

$$J = \begin{bmatrix} B \ , \ I = 0 \\ C \ , \ I = 1 \end{bmatrix}$$

of a global or regional risk occurring at the future time point

$$T,$$

as a random contraction of the random variable

$$J = \begin{bmatrix} B \ , \ I = 0 \\ C \ , \ I = 1 \end{bmatrix}$$

via the random variable

$$W = e^{-\delta T}$$

provides analysts, modelers, decision makers and other experts with valuable probabilistic information for proactive thinking and acting in the context of global or regional risk governance. More precisely, the indirect evaluation of the probability density function

$$f_U(\upsilon)$$

of the random contraction

$$U = Je^{-\delta T},$$

through the characteristic function

$$\varphi_U(u) = \int_0^1 [(1-\rho)\varphi_B(uw) + \rho P_N(P_L(\varphi_X(uw)))]d\left[1 - F_T\left(-\frac{1}{\delta}logw\right)\right]$$

gives to the global risk governance experts the extremely important opportunity to thoroughly and proactively investigate the impact of the severity

$$J = \begin{bmatrix} B\ , I = 0 \\ \\ C\ , I = 1 \end{bmatrix}$$

of a global or regional risk occurring at the future random time point

$$T.$$

It is readily understood that such an investigation strongly supports strategic thinking and acting for the successful treatment of a global risk before its occurrence at the future random time point

$$T.$$

It is of particular practical importance to clarify the interconnection between the existence of a most probable and best scenario for the global or regional risk severity

$$J = \begin{bmatrix} B\ , I = 0 \\ \\ C\ , I = 1 \end{bmatrix}$$

and the existence of a most probable and best scenario for the discounted global or regional risk severity

$$U = Je^{-\delta T},$$

It is known that the presence of a unique mode at the point

$$\beta = 0$$

for the probability density function

$$f_B(\beta)$$

of the positive random variable

$$B$$

and the presence of unique mode at the point

$$c = 0$$

for the probability density function

$$f_C(c)$$

of the positive random sum

$$C = X_1 + X_2 + ... + X_S$$

implies the presence of unique mode at the point

$$j = 0$$

for the probability density function

$$f_J(j)$$

of the random variable

$$J = \begin{bmatrix} B \; , \; I = 0 \\ \\ C \; , \; I = 1 \end{bmatrix}$$

and consequently the presence of a unique mode at the point

$$\upsilon = 0$$

for the probability density function

$$f_U(\upsilon)$$

of the random contraction

$$U = Je^{-\delta T}$$

[26]. Equivalently, the existence of a most probable and best scenario for the global risk severity

$$J = \begin{bmatrix} B \ , \ I = 0 \\ \\ C \ , \ I = 1 \end{bmatrix}$$

implies the existence of a most probable and best scenario for the discounted global or regional risk severity

$$U = Je^{-\delta T}.$$

Moreover, it is of some particular practical importance to clarify the role of the random variable

$$T$$

and the random variable

$$W = e^{-\delta T}$$

in the probabilistic behaviour of the random contraction

$$U = Je^{-\delta T}.$$

We suppose that the probability density function

$$f_T(t)$$

of the random variable

$$T$$

satisfies the conditions for making the probability density function

$$f_W(w)$$

of the random variable

$$W = e^{-\delta T}$$

unimodal at the point

$$w = 0$$

In this case the random contraction

$$U = Je^{-\delta T}$$

has a probability density function

$$f_U(\upsilon)$$

with a unique mode at the point

$$\upsilon = 0$$

[26]. Equivalently, a most probable and best scenario exists for the discounted global or regional severity

$$U = Je^{-\delta T}$$

based only on the presence of a unique mode at the point

$$w = 0$$

for the probability density function

$$f_W(w)$$

of the random variable

$$W = e^{-\delta T}.$$

Consequently, it can be said that the role of the random variable

$$T$$

and the role of the random variable

$$W = e^{-\delta T}$$

are important for thinking, deciding and acting proactively in the context of global or regional risk governance.

For example, if the random variable

$$T$$

follows the exponential distribution with parameter

$$\mu$$

satisfying

$$\mu < \delta$$

then the discounted global or regional risk severity

$$U = Je^{-\delta T}$$

is a α -unimodal contraction with probability density function

$$f_U(\upsilon)$$

having a unique mode at the point

$$\upsilon = 0,$$

where

$$\alpha = \mu/\delta.$$

Below we provide an assumption which extends the practical applicability of the contribution of the present section.

The contribution of the present section consists of the interpretation of the random variable

$$J = \begin{bmatrix} B \ , I = 0 \\ \\ C \ , I = 1 \end{bmatrix}$$

as the severity of a global or regional risk occurring at the future time point

$$T$$

and the interpretation of the random variable

$$U = Je^{-\delta T}$$

as the present value, as viewed from the time point 0, of the global or regional risk severity

$$J = \begin{cases} B \ , \ I = 0 \\ C \ , \ I = 1 \end{cases}$$

arising at the future random time point

$$T.$$

It is readily understood that the contribution of the present section is substantially extended by making the assumption that the random variable

$$B$$

denotes the severity of a risk with its cause being of a low intensity and the random variable

$$C = X_1 + X_2 + ... + X_S.$$

denotes the severity of the same risk with its cause being of a high intensity.

4.16 Model of Global or Regional Risk Severity Reduction

The present section concentrates on the interpretation of the random variable

$$U = JW$$

as a stochastic model for the severity of a global or regional risk after the application of a global or regional risk severity reduction operation.

We suppose that a risk occurs at the future random time point

$$T.$$

The risk is global with probability

$$\rho$$

and the risk is regional with probability

$$1 - \rho.$$

We also suppose that the discrete random variable

$$N$$

denotes the number of countries affected by the occurrence of the global risk and the discrete random variable

$$L_n$$

denotes the number of the impaired systems of the

$$n\text{th}$$

country because of the occurrence of the global risk. Hence the discrete random variable

$$S = L_1 + L_2 + \ldots + L_N$$

denotes the number of the impaired systems of the

$$N$$

countries because of the occurrence of the global risk. In addition, we suppose that the positive random variable

$$X_s$$

denotes the severity of the economic loss of the

$$s\text{th}$$

impaired system. Hence the positive random variable

$$C = X_1 + X_2 + \ldots + X_S.$$

denotes the severity of the global risk occurring at the future random time point

$$T$$

with probability

$$\rho.$$

Moreover, we suppose that the positive random variable

$$B$$

denotes the severity of the regional risk occurring at the future random time point

$$T$$

with probability

$$1 - \rho.$$

If

$$I$$

is a Bernoulli random variable with probability function

$$P(I = 0) = 1 - \rho,$$

$$P(I = 1) = \rho$$

then the random variable

$$J = \begin{cases} B, & I = 0 \\ C, & I = 1 \end{cases}$$

denotes the severity of the global or regional risk occurring at the future random time point

$$T.$$

Furthermore, we suppose that the random variable

$$W$$

with values in the interval

$$(0, 1),$$

denotes the impact of applying a global or regional risk severity reduction operation to the conditions and the cause of the global or regional risk. Under the above assumptions, the random variable

$$U = JW$$

can be interpreted as the severity of the global or regional risk after the application of a global or regional risk severity reduction operation to such risks.

Below, we provide an assumption which substantially extends the practical applicability of the present section.

The contribution of the present section consists of the interpretation of the random variable

$$J = \begin{cases} B\ ,\ I = 0 \\ C\ ,\ I = 1 \end{cases}$$

as the severity of a global or regional risk occurring at the future random time point

$$T$$

and the interpretation of the random variable

$$U = JW$$

as the severity of a global or regional risk after the application of a global or regional risk severity reduction operation to such risks. It is readily understood that the contribution of the present section can be extended by making the assumption that the random variable

$$B$$

denotes the severity of a risk with its cause being of a low intensity and the random variable

$$C = X_1 + X_2 + ... + X_S$$

denotes the severity of the same risk with its cause being of a high intensity.

Furthermore, we suppose that the discrete random variable

$$N$$

has probability generating function

$$P_N(z),$$

the discrete random variables of the sequence

$$\{L_n n = 1, 2, ...\},$$

are independent and distributed as the random variable

$$L$$

with probability generating function

$$P_L(z),$$

the positive random variables of the sequence

$$\{X_s, s = 1, 2, ...\},$$

are independent and distributed as the random variable

$$X$$

with characteristic function

$$\varphi_X(u),$$

the positive random variable

$$B$$

has characteristic function

$$\varphi_B(u),$$

the Bernoulli random variable

$$I$$

has probability generating function

$$P_I(z) = 1 - \rho + \rho z$$

and the random variable

$$W$$

with values in the interval

$$(0, 1),$$

has distribution function

$$F_W(w).$$

If

$$N,$$

$$\{L_n, n = 1, 2, ...\},$$

$$\{X_s, s = 1, 2, ...\}$$

$$B,$$

$$I$$

and

$$W$$

are independent then *Theorem* 3.8.1 implies that the random variable

$$U = JW$$

is a random contraction of the random variable

$$J = \begin{cases} B \ , \ I = 0 \\ C \ , \ I = 1 \end{cases}$$

via the random variable

$$W$$

and *Theorem* 3.9.1 implies that the characteristic function of the random variable

$$U = JW$$

has the form

$$\varphi_U(u) = \int_0^1 [(1-\rho)\varphi_B(uw) + \rho P_N(P_L(\varphi_X(uw)))] dF_W(w).$$

In the following we make clear that the representation of the random variable

$$U = JW$$

as a random contraction of the random variable

$$J = \begin{bmatrix} B \text{ , } I = 0 \\ \\ C \text{ , } I = 1 \end{bmatrix}$$

via the random variable

$$W$$

substantially extends the practical applicability in the context of global or regional risk governance of the interpretation of the random variable

$$U = JW$$

as the severity of a global or regional risk after the application of a risk severity reduction operation to such risks. More precisely, the representation of the random variable

$$U = JW$$

as a random contraction of the random variable

$$J = \begin{bmatrix} B \text{ , } I = 0 \\ \\ C \text{ , } I = 1 \end{bmatrix}$$

via the random variable

$$W$$

makes possible the explicit analytical evaluation of the characteristic function

$$\varphi_U(u) = \int_0^1 [(1-\rho)\varphi_B(uw) + \rho P_N(P_L(\varphi_X(uw)))]dF_W(w).$$

Since an explicit analytical and direct evaluation of the probability density function

$$f_U(v)$$

corresponding to the random contraction

$$U = JW$$

is not possible then the investigation of the probabilistic behaviour of the random contraction

$$U = JW$$

is implemented by making use of the characteristic function

$$\varphi_U(u) = \int_0^1 [(1-\rho)\varphi_B(uw) + \rho P_N(P_L(\varphi_X(uw)))]dF_W(w).$$

In particular, we make use of the above characteristic function, the inversion theorem for characteristic functions and the Fast Fourier Transform algorithm in order to obtain the probability density function

$$f_U(\upsilon)$$

of the random contraction

$$U = JW$$

denoting the severity of a global or regional risk after the application of a risk severity reduction operation to such risks.

In other words, the representation of the random variable

$$U = JW,$$

denoting the severity of a global or regional risk after the application of a global or regional risk severity reduction operation to such risks, as a random contraction of the random variable

$$J = \begin{bmatrix} B \ , \ I = 0 \\ \\ C \ , \ I = 1 \end{bmatrix}$$

via the random variable

$$W$$

provides analysts, modelers, decision makers and other experts with valuable probabilistic information for proactive treatment of global or regional risks. More precisely, the indirect evaluation of the probability density function

$$f_U(\upsilon)$$

of the random contraction

$$U = JW$$

through the characteristic function

$$\varphi_U(u) = \int_0^1 [(1-\rho)\varphi_B(uw) + \rho P_N(P_L(\varphi_X(uw)))]dF_W(w)$$

gives to the global or regional risk governance experts the extremely important opportunity to proactively reduce the severity

$$J = \begin{cases} B\ , & I = 0 \\ C\ , & I = 1 \end{cases}$$

of a global or regional risk before its occurrence. More precisely, the contribution of the present section can be extended by combining the random contraction

$$U = JW,$$

denoting the severity of a global or regional risk after the application of a global or regional risk severity reduction operation to such risks, in the random time interval

$$[0, T)$$

where

$$T$$

is the occurrence future random time point of a global or regional risk. It is easily seen that such a combination gives the significant advantage to global or regional risk governance experts to make use of the probabilistic information provided by the random contraction

$$U = JW$$

for developing, assessing, selecting and implementing global or regional risk severity reduction operations for proactive treatment in the random time interval

$$[0, T)$$

of a global or regional risk occurring at the future random time point

$$T.$$

It is of particular practical importance to clarify the interconnection between the existence of a most probable and best scenario for the global or regional risk severity

$$J = \begin{cases} B\ , & I = 0 \\ C\ , & I = 1 \end{cases}$$

and the existence of a most probable and best scenario for the severity

$$U = JW$$

of a global or regional risk after the application of a global or regional risk severity reduction operation to such risks. It is known that the presence of a unique mode at the point

$$\beta = 0$$

for the probability density function

$$f_B(\beta)$$

of the positive random variable

$$B$$

and the presence of unique mode at the point

$$c = 0$$

for the probability density function

$$f_C(c)$$

of the random sum

$$C = X_1 + X_2 + ... + X_S$$

implies the presence of unique mode at the point

$$j = 0$$

for the probability density function

$$f_J(j)$$

of the random variable

$$J = \begin{cases} B\ , \ I = 0 \\ C\ , \ I = 1 \end{cases}$$

and consequently the presence of a unique mode at the point

$$\upsilon = 0$$

for the probability density function

$$f_U(\upsilon)$$

of the random contraction

$$U = JW$$

[26]. Equivalently, the existence of a most probable and best scenario for the global or regional risk severity

$$J = \begin{cases} B\ , \ I = 0 \\ C\ , \ I = 1 \end{cases}$$

implies the existence of a most probable and best scenario for the severity

$$U = JW$$

of a global or regional risk after the application of a global or regional risk severity reduction operation to such risks.

Moreover, it is of particular practical importance to provide evidence concerning the role of the random variable

$$W$$

in the probabilistic behaviour of the random contraction

$$U = JW$$

It is known that the presence of a unique mode at the point

$$w = 0$$

for the probability density function

$$f_W(w)$$

of the random variable

$$W$$

implies the presence of a unique mode at the point

$$\upsilon = 0$$

for the probability density function

$$f_U(\upsilon)$$

of the random contraction

$$U = JW$$

[26]. Equivalently, a most probable and best scenario exists for the global or regional risk severity

$$U = JW,$$

after the application of a global or regional risk severity reduction operation to such risks, based only on the presence of a unique mode at the point

$$w = 0$$

for the probability density function

$$f_W(w)$$

of the random variable

$$W.$$

Consequently, it can be said that the role of the random variable

$$W$$

is important in selecting and implementing a global risk severity reduction operation. More clearly, the selection and implementation of an efficient global risk severity reduction operation can substantially contribute to a proactive and successful treatment of a global risk. Below, we consider two interesting cases for the probability density function

$$f_W(w)$$

of the random variable

$$W.$$

We suppose that the random variable

$$W$$

follows the beta distribution with probability density function

$$f_W(w) = aw^{a-1}, a > 0, 0 < w < 1$$

which has a unique mode at the point

$$w = 0$$

under the assumption that

$$0 < a \leq 1$$

It is known that under the above assumption the random variable

$$U = JW$$

is a α -unimodal random contraction with probability density function

$$f_U(\upsilon)$$

having a unique mode at the point

$$\upsilon = 0$$

[26]. Consequently, if the random variable

$$W,$$

with probability density function

$$f_W(w) = aw^{a-1}, a > 0, 0 < w < 1$$

denotes the impact of a global or regional risk severity reduction operation applied to the cause and the conditions of a global or regional risk then a most probable and best scenario exists for the severity

$$U = JW$$

of such risks.

Moreover, we suppose that the random variable

$$W$$

follows the beta distribution with probability density function

$$f_W(w) = \nu(1 - w)^{\nu-1}, \nu > 0, 0 < w < 1,$$

which has a unique mode at the point

$$w = 0$$

under the assumption that

$$\nu \geq 1.$$

It is known that under the above assumption the random variable

$$U = JW$$

is a υ -unimodal random contraction with probability density function

$$f_U(\upsilon)$$

having a unique mode at the point

$$\upsilon = 0$$

[26]. Consequently, if the random variable

$$W,$$

following the beta distribution with probability density function

$$f_W(w) = \nu(1 - w)^{\nu-1}, \nu \geq 1, 0 < w < 1,$$

denotes the impact of a global or regional risk severity reduction operation applied to the cause and the conditions of a global or regional risk then a most probable and best scenario exists for the severity

$$U = JW$$

of such risks.

4.17 Conclusions

It constitutes a direct and indisputable recognition that the practical contribution of the present chapter mainly consists of the very useful incorporation of various well known and particularly significant stochastic models such as.

- *Discrete Random Sum*
- *Positive Random Sum*
- *Present Value of Positive Random Sum*
- *Positive Bernoulli Selecting Process*
- *Present Value of Positive Bernoulli Selecting Process*
- *Random Contraction of Positive Bernoulli Selecting Process*

for the precise definition, the thorough investigation and the efficient application of fundamental concepts such as.

- *Number of Impaired Systems*
- *Global Risk Severity*
- *Global or Regional Risk*

arising in the context of global risk governance. In addition, the above stochastic models are readily adopted as particularly significant for the reliable description, the optimal selection and the successful implementation of operations such as.

- *Discounting Global Risk Severity*
- *Reducing Global Risk Severity*
- *Discounting Global or Regional Risk Severity*
- *Reducing Global or Regional Risk Severity.*

also arising in the context of global risk governance. Moreover, the practical contribution of the present chapter is widely revealed by making quite clear the very useful role for the context of global risk governance, the presence of the properties of unimodality, stability, selfdecomposability and infinite divisibility in the probability distributions corresponding to the previously mentioned stochastic models. In particular, it is shown that the presence of the property of ordinary unimodality or equivalently the presence of a unique mode at the point 0 for such probability distributions imply the existence of a most probable and best scenario for concepts such as number of impaired systems, global risk severity, global or regional risk severity and operations such as discounting global risk severity, reducing global risk severity, discounting global or regional risk severity, reducing global or regional risk severity.

It can be said that these concepts and operations are generally recognized as very strong tools of global risk governance for the description, investigation and solution of problems with particular practical importance.

Global risks are defined in this study as having a global geographical scope, an inter-industrial presence, and exceptionally critical stages of economic and social participation that necessitate a major multi-stakeholder input. In addition, global risks demand an extremely extensive priority in decision making allowance. This work can be seen as a contribution to the area of ongoing global risks and the area of forthcoming global risks, particularly necessary for the implementation of very important interdisciplinary research activities. The theoretical and practical results of this work are strongly connected to several quite useful factors. The present work mainly concentrates on the contribution of probability theory in the advancement of the practical applicability of global risk governance. More precisely, the work introduces structural stochastic concepts and fundamental stochastic results for the formulation of stochastic models of various global risk governance operations particularly valuable in proactive treatment of several groups of global risks.

References

1. A.C. Cullen, H.C. Frey, *Probabilistic Techniques in Exposure Assessments: A Handbook for Dealing with Variability and Uncertainty in Models and Inputs* (Plenum Press, New York, 1999)
2. M. Pinsky, S. Karlin, *An Introduction to Stochastic Modeling*, 4th edn. (Elsevier Inc., Oxford, UK, 2011)
3. R. Lidskog, L. Soneryd, Y. Uggla, *Transboundary Risk Governance* (Earthscan Publications Ltd., London, 2010)
4. L. Bayer, J., Lofsedt, R. and Sjostedt, G. (eds.), *Transboundary Risk Management* (Earthscan Publications Ltd., London, 2001)
5. J. Tait, A. Bruce, (2004) Global change and transboundary risks, in *Risk Analysis and Society: An Interdisciplinary Characterization of the Field*, ed. by T. McDaniels, M. Small (Cambridge University Press, Cambridge), pp. 367–419. (Commissioned by Society for Risk Analysis for the International Symposium on Risk and Governance, Warrenton, VA, USA, June 2000).
6. P. Grossi, H. Kunreuther (eds.), *Catastrophe Modeling: A New Approach to Managing Risk* (Springer, New York, 2005)
7. A. Posner, *Catastrophe: Risk and Response* (Oxford University Press, Oxford, 2005)
8. C. Lyall, J. Tait (eds.), *New Modes of Governance: Developing an Integrated Policy to Science, Technology, Risk and the Environment*, (Ashgate Publishing Ltd, Aldershot, Hampshire, 2005)
9. N. Bostrom, M. Cirkovic, Introduction, ed. by N. Bostrom, M. Cirkovic Global Catastrophic Risks (Oxford University Press, Oxford, 2008), pp. 1–29
10. C. Bunting, O. Renn, M.V. Florin, R. Cantor. Introduction to the IRGC risk governance framework,in *The John Liner Review*, vol. 21 (2007), pp. 7–26
11. O. Renn, (2008) White paper on risk governance: towards an integrative framework, in *Global Risk Governance: Concept and Practice Using the IRGC Framework*, vol. 1, ed. by O. Renn, Walker (International Risk Governance Council Bookseries, Springer, Dordrecht), pp. 3–73
12. M. Ghertman, J. Obadia, J.-L. Arregle (eds.), *Statistical Models for Strategic Management* (Springer, New York, 2004)
13. A. Henry, *Understanding Strategic Management* (Oxford University Press, Oxford, 2007)

14. H. Tijms, *Stochastic Modeling and Analysis: A Computational Approach*, vol. 13 (Wiley, New York 1988), pp. 1–3
15. P.T. Artikis, Stochastic concepts facilitating strategic thinking in global risk governance. J. Stat. Manag. Syst. **17**(2), 183–194 (2014)
16. N. Johnson, S. Kotz, A. Kemp, *Univariate Discrete Distributions*, 3rd edn. (Wiley, New York, 2005)
17. M. Yamazato, Unimodality of infinitely divisible distribution functions of class L. Ann. Probab. **6**, 523–531 (1978)
18. P. Medgyessy, On the unimodality of discrete distributions. Period. Math. Hung. **2**, 245–257 (1972)
19. P.T. Artikis, C.T. Artikis, K.A. Agorastos, A. Vlachos, Stochastic derivation and application in information risk frequency reduction of a class of discrete random variables. J. Stat. Manag. Syst. **12**, 59–64 (2009)
20. N. Krishnaji, A characteristic property of the yule distribution. Sankya A **32**, 343–346 (1970)
21. P.T. Artikis, C.T. Artikis, C.E. Foundas, P. Hatzopoulos, Discrete renewal and selfdecomposable distributions in modeling information risk management operations. J. Stat. Manag. Syst. **9**, 73–85 (2006)
22. M. Gomes, D. Pestana, The use of fractional calculus in probability. Portugaliae Mathematica **37**, 259–271 (1978)
23. R. Olshen, S. Savage, A generalized unimodality. J. Appl. Probab. **7**, 21–34 (1970)
24. F. Steutel, K. van Harn, *Infinite Divisibility of Probability Distributions on the Real Line* (Marcel Dekker Inc., New York, 2004)
25. E. Lukacs, *Characteristic Functions* (Hafner, New York, 1970)
26. P. Medgyessy, On a new class of unimodal infinitely divisible distribution functions and related topics. Studia Sci. Math. Hungar. **2**, 441–446 (1967)